Integrated
Virus Detection

Integrated
Virus Detection

Charles H. Wick

CRC Press
Taylor & Francis Group
Boca Raton London New York

CRC Press is an imprint of the
Taylor & Francis Group, an **informa** business

CRC Press
Taylor & Francis Group
6000 Broken Sound Parkway NW, Suite 300
Boca Raton, FL 33487-2742

First issued in paperback 2020

ISBN-13: 978-1-4822-3005-5 (hbk)
ISBN-13: 978-0-367-65893-9 (pbk)

Library of Congress Cataloging-in-Publication Data

Wick, Charles Harold, author.
  Integrated virus detection / Charles H. Wick.
  p. ; cm.
  Includes bibliographical references and index.
  ISBN 978-1-4822-3005-5 (hardcover : alk. paper)
  I. Title.
  [DNLM: 1. Viruses--isolation & purification--Laboratory Manuals. 2. Virology--methods--Laboratory Manuals. 3. Virus Diseases--diagnosis--Laboratory Manuals. 4. Virus Physiological Phenomena--Laboratory Manuals. QW 25]

QR201.V55
616.9'101--dc23   2014039068

Visit the Taylor & Francis Web site at
http://www.taylorandfrancis.com

and the CRC Press Web site at
http://www.crcpress.com

This work is dedicated to
Virginia Mason Morris Wick

Contents

Abbreviations and Glossary

Bacteriophage: Virus that attacks bacteria

CPC: Condensate particle counter

Cross flow: Sweeping action created by fluid flow across a membrane

Dalton (D): Molecular weight of macromolecule

Diafiltration: Dialysis type of filtration

Dialysis: Separation of salts and microsolutes from macromolecular solutions by concentration gradient

DMA: Differential mobility analyzer

Electrospray (ES): Atomizing a liquid by injecting across an electrical potential

GEMMA: Gas-phase electrophoretic mobility molecular analyzer

IVDS: Integrated Virus Detection System

MS2: Type of bacteriophage

MW: Molecular weight

MWCO: Molecular weight cut-off

Phage: *See* bacteriophage

Plaque-forming unit (pfu): Quantity of virus required to form viable colony

SMPS: Scanning mobility particle sizer

T7: Type of bacteriophage

Tangential flow: *See* cross flow

Ultrafiltration (UF): Separation of salts and microsolutes from macromolecular solutions by hydrostatic pressure

Preface

Viruses are dangerous and are waiting for the chance to break out and infect society. Humankind lives in uncertain times, and this is our future without vigilant and effective safeguards. This is further exasperated by the ability to create new viruses, from scratch, within a few weeks. These contagions have the potential to escape confinement, infect people, and harm the plants, insects, and animals we depend upon to survive. Society is in danger, but we no longer need to wait for an innovative method to detect these malicious microbes.

Many viruses can be detected simultaneously using the Integrated Virus Detection System (IVDS). *Integrated Virus Detection* describes this technology and provides many examples, including a chapter on previously unknown viruses found in honeybees that includes charts on seasonal and yearly variations. This straightforward technology can be used to detect previously unknown viruses collected from environmental and other complex biological matrices.

This book presents in a single reading a summary of more than 10 US patents that were issued for the invention of the Integrated Virus Detection System. IVDS is powering a revolution in humankind's ability to rapidly detect, measure, and monitor viruses and virus-like particles. Three facts make this possible: (1) viruses are in the size range of 20–800 nm; (2) there is sufficient disparity in each virus species' particle size such that size data may be used for detection and preliminary identification; and (3) virus particle density is distinct from other nanoparticles. IVDS is an innovative, non-chemical approach to near real-time comprehensive screening of all viruses present in a sample.

The entire virus-screening process from sample collection to review of results can occur within a few minutes. The absence of virion-particles is compelling evidence of a true-negative. Large numbers of samples may be processed to provide an excellent means to pre-screen samples for judicious targeting of subsequent tests such as mass spectrometry proteomics and polymerase chain reaction (PCR). The IVDS system employs relatively simple, low cost equipment, with few reagents. New advances indicate that the entire virion-size range can be measured concurrently and the process automated thus making virus detection simple and quick.

The IVDS technique is not limited by historical concerns and has been used successfully for many years to monitor the more than 20 virus species common in honeybees. Without effective viral detection, our environment is in danger. IVDS offers the solutions we need to allow for the successful treatment of apiaries expressing high viral loading and return the honeybees to health.

This book follows the invention of the IVDS technology, the development of the first instruments, and the use of IVDS methods to common applications. The reader can expect to gain a working understanding of the IVDS instrument and how to use the technology and methods in several applications.

Acknowledgments

I would like to thank those who made this book possible: Patrick McCubbin for his faithful service and for his humor and dedication that were both welcome and rewarding, and Dr. Michael F. Stanford for his steadfast enthusiasm for keeping me on track to finish this book.

There are many to thank in support for the honeybee work. One of these hardworking people is David A. Wick, president of BVS, Inc., whose dedication and perseverance helped take IVDS to the field. Special thanks are extended to enthusiastic beekeepers everywhere and their local, state and national associations, the Almond Board of California, Project Apis m., the US Department of Agriculture, the University of Montana, Montana State University, and Bee Alert Technology, Inc.

In particular, I appreciate the work done by my son Harrison S. Wick for proofreading, formatting, and organizing the manuscript and final draft, and by my daughter Jinnie M. Wick for her graphic design and placing all the figures into proper form.

About the Author

Charles H. Wick, PhD, is a retired senior scientist from the US Army Edgewood Chemical Biological Center (ECBC) where he served both as a manager and a research physical scientist and made significant contributions to forensic science. Although his 40-year professional career has spanned both the public and military sectors, his best-known work in the area of forensic science was done in concert with the Department of Defense (DoD).

After earning four degrees including the PhD from the University of Washington, Dr. Wick worked in civilian occupations in the private sector for 12 years inventing, which led to a patent, writing numerous publications, and gaining international recognition among his colleagues.

In 1983, Dr. Wick joined the Vulnerability/Lethality Division of the US Army Ballistic Research Laboratory, where he quickly achieved recognition as a manager and principal investigator. At this point, he made one of his first major contributions to forensic science and to the field of antiterrorism. His team was the first to utilize then-current technology to model sub-lethal chemical, biological, and nuclear agents. This achievement was beneficial to all areas of the DoD and also to the North Atlantic Treaty Organization (NATO), and gained Dr. Wick international acclaim as an authority on individual performance for operations conducted on a nuclear, biological, and chemical (NBC) battlefield.

During his career in the US Army, Dr. Wick rose to the rank of lieutenant colonel in the Chemical Corps. He served as a unit commander for several rotations, a staff officer for 6 years (division chemical staff officer for two rotations), deputy program director of Biological Defense Systems, and retired as commander of the 485th Chemical Battalion in April 1999.

Dr. Wick continued to work for DoD as a civilian at ECBC. Notable achievements, one of which earned him the Department of the Army Research and Development Award for Technical Excellence and a Federal Laboratory Consortium Technology Transfer Award in 2002, include his involvement in the invention of the Integrated Virus Detection System (IVDS), a fast-acting, highly portable, user-friendly, extremely accurate and efficient system for detecting, screening, identifying, and characterizing viruses.

The IVDS can detect and identify the full spectrum of known, unknown, and mutated viruses, from AIDS to hoof-and-mouth disease, to West Nile virus, and beyond. This system is compact, portable, and does not rely upon elaborate chemistry. The second award-winning achievement was his leadership in inventing a method for detecting microbes using mass spectrometry proteomics. Each of these projects represents a determined 10-year effort and a novel approach for detecting and classifying microbes from complex matrices.

Throughout his career, Dr. Wick made lasting and important contributions to forensic science and antiterrorism. He holds several US patents in the area of microbe detection and classification. He has written more than 45 civilian and military publications and received myriad awards and citations including the Department of the Army Meritorious Civilian Service Medal, the Department of the Army Superior Civilian Service Award, two Army Achievement Medals for Civilian Service, a Commander's Award for Civilian Service, the Technical Cooperation Achievement Award, and 25 other decorations and awards for military and community service.

1 Detection of Whole Viruses

1.1 INTRODUCTION

A new method for the detection and characterization of viruses has been indicated and needed for years. This situation has reached and continues to soar to new alarm status with the rapidly expanding number of new and emerging viruses that tax present detection methods. Adding the cost of current methods to analyze large numbers of samples and a clear and urgent situation has developed. Fortunately, a new means for detecting the present or absence of viruses has been developed that capitalizes on the fundamental physical properties associated with these tiny microbes.

The Integrated Virus Detection System (IVDS) utilizes exquisite but regular methods to purify and concentrate samples for sizing and counting using the usual methods of differential mobility analysis and condensation particle counting. Thus a new method that is essentially a virus particle counter has been invented, patented, demonstrated, and placed into commercial use for a wide range of both enveloped and non-enveloped viruses.

This invention was possible because of two major features of viruses: (1) different families have different sizes and (2) viruses have physical properties different from those of other submicron-sized particles in the environment. The first feature has exceptions, e.g., the polymorphs, but size was found not to be a limitation because of the other features and scarcity of polymorphs in most environments. The second feature known as the virus window is exceptionally useful in that the virus families, particularly those pathogenic to humans, can be separated by their physical features. These features allowed the isolation and characterization of many viruses in the natural environment as demonstrated by the long-time monitoring of viruses in honeybees.

This book describes the story of this invention, the various enhancements to the technology, and some of the applications. The story starts in the middle 1990s following the First Gulf War and the then urgent need for a means to detect microbes, particularly viruses. The new method needed to be better and more useful than the molecular biological processes in use then. It also had to improve on life cycle costs and simplified logistics. This translates to three features for the new system: (1) no complex chemistry or biochemical reactions or features; (2) no shelf life issues; and (3) the ability to detect all the viruses in a sample in a single pass without error. False negatives would not be tolerated; likewise false positives are just as useless. This guidance was the beginning of the story.

I was very familiar with systems that were proposed and rushed forward for use. The systems were mostly cosmetic in nature and generally useless. I tried several of them that were being rushed toward use and they lacked all three pillars of the guidance. It was a terrific experience to have both the on-hands field experience with these systems as a soldier and the opportunity to work out a solution as a scientist. In summary, everyone involved was in a rush to produce a method but few biologists were assigned and fewer yet with cross training were involved in the effort. It was bemusing and extremely frustrating to have a project slowed down by chemists and engineers who approached the biological detection problem from their views and always thought they could figure out an answer if they had enough funding.

Many participants I appreciated and admired, including some chemists and engineers, were optimistic and full of enthusiasm. Their support helped sustain my enthusiasm and determination that resulted in the creation of this new instrument. I am sure this is the normal experience of any scientist who works outside the well-traveled corporate path, but that is the subject of another book that will probably be much more exciting to read.

I would say that the key to success is perseverance: staying focused on the goal. Help and support for a good idea will emerge and eventually success will follow. Two key supporters, DARPA and the Program Manager for Biological Defense, appeared and with their help the first instruments were developed and tested.

The IVDS method for detecting viruses changes all aspects of virus analysis. It requires no culturing or no reagents, provides quick analysis, and is generic to all viruses. It has a long shelf life and can be stored until needed. It counts whole virus particles, is suitable for counting all types, and thus can also count unknown viruses and previously undetectable viruses.

The beta instrument was used to examine viruses separated from complex media with a sensitivity that approached 10^4 viruses per milliliter—a number greatly improved on new instruments. Other virus characteristics were observed and calibrated, e.g., their passage through filter membranes and the unexpected survival of MS2 in both extreme temperature and pH environments. The 4-nanometer separation of viruses of similar size aids in this direct structural characterization of large numbers of viruses.

One of the first tests of this new technology was to compare it with other methods such as small angle neutron scattering (SANS). Results were comparable (Kuzmanovic et al. 2003). A second objective was to demonstrate the effectiveness of IVDS in counting viruses under a wide range of conditions. This is important as the many inherent challenges to virus detection and analysis include purification and concentration from background material. This was accomplished for several viruses from many environmental samples of various soils including sand, drinking water, seawater, and plants.

These early results were all improved upon over the following 10 years. The second (and first commercial) version is still in use and proved to be a rugged and reliable device (see Chapter 10 about application to monitoring honeybee viruses). The future instruments are proposed as small portable, automatic devices that provide greater resolution. We expect to see them in local drug stores as self-testing instruments. The instrument details are covered in Chapter 2.

1.2 HISTORICAL BACKGROUND

1.2.1 Introduction

The ability to detect virus particles rapidly was a vexing issue to both the military and public health communities in the 1990s. The need for a rapid and trustworthy virus detector remained urgent for all levels of decision makers who required assurances of early warnings from virus threats. A new method was indicated and the need led to the development of the IVDS.

Working with viruses is not an easy job. It is interesting to note that the need or rather the requirement for the detection of viruses was identified in the 1930s and only now can virologists see a device that can undertake this exacting and difficult task—the new IVDS methodology.

The IVDS utilizes a physical process to detect viruses. This is a departure from the traditional approaches that historically utilized obscure chemical reactions, complicated reagents, and exacting and difficult procedures for detection. The IVDS is based in four physical stages: (1) collection, (2) separation, (3) purification, and (4) detection. Development and risks were considered and these early issues are discussed in this section as a matter of completeness. The first IVDS development schedule and cost estimates suggested that an advance prototype could be made in 2 years. The IVDS system promised at the beginning a very sensitive, broad spectrum generic virus detector free from reagent-based reactions, simple to operate, and allowing continuous monitoring and recording of virus levels outdoors.

1.2.2 Need for New Detection System

In a very real sense, viruses have already demonstrated their effectiveness as weapons of warfare through a number of historic events. The World War I pandemic of Spanish Influenza in 1918–1919 killed approximately 20 million people and is considered by many to have been a primary factor in the defeat of the German Army. According to virologist Robert Webster (1993, p. 38), "It wasn't the Americans coming to Europe, it was the virus they brought that did the job." Similarly, epidemics caused by pathogens, particularly smallpox, brought from Europe to America wiped out nearly 90% of the Native American population, and there is clear evidence that smallpox was deliberately used as a weapon according to Henry Bouquet as early as 1763 (Bouquet 1763).

In 430 BCE, sequestering of the Athenian population behind walls as a defensive move during the Peloponnesian War caused a great epidemic, probably smallpox or a virulent strain of measles (Temin 1993). Armies of ancient Greece, and later those of Rome and Persia were known to poison the drinking water of their enemies with diseased remains (Stockholm International Peace Institute 1971). Returning to more recent developments, viral agent candidates tested by the Japanese biological warfare program in the 1930s and 1940s included yellow fever, hepatitis, encephalitis, and hemorrhagic fever (Harris 1992). Most recently, it has been suggested that prolonging the Desert Storm or the Panamanian actions by even a few weeks could have exposed US troops to serious threats of Crimean-Congo hemorrhagic fever or sand fly fever viruses, respectively (Morse 1993).

The first detection issue was defining the problem of monitoring, particularly airborne viruses, from the perspective of someone who must accomplish this task while facing the inherent problems of determining which methodologies will actually work and devising workable solutions. The solution called for semi-quantitative analysis of background interference, sensitivities, specificities, tolerances, cost factors, and design specifications. The approach had to differ fundamentally from past practices espousing a particular method (often in a highly idealized and irrelevant setting such as a clean room that did not consider the outdoor bioaerosol issues or the contamination inherent in a messy environment) or more generally, qualitative issues surrounding viral warfare.

1.2.3 STATEMENT OF PROBLEM

Airborne virus detection presents many inherent challenges. From the standpoint of defense against military and terrorist threats and the more global perspective of public health, the detection of airborne viruses is a particular technological challenge that makes chemical warfare (CW) detection look easy in comparison (Wick et al. 1995; Wick et al. 1997a; Wick et al. 1997b). Factors contributing to the extraordinary difficulty of the task include the wide range of viruses carried in the air, the exceedingly large background of biological components relative to infectious levels of viruses, and the problem of mutation by both natural and artificial means that greatly enlarged the inventory of potential biological warfare viral weaponry.

Purification and concentration of the background material are required, whatever the detection method, to be used in subsequent steps. Background loading of biological samples is an astronomically large undertaking, and the virus materials of interest must be separated from the biological debris. A five-millionfold concentration factor is required, whether the next step is based on biological activity or physical attributes. There is no purpose in even considering the detection of viruses until the purification and concentration problems are resolved (Wick et al. 1998a).

1.2.4 HISTORIC METHODS

This section will discuss the historic methods for the purification, extraction, and concentration of viruses. For more than 70 years, scientists have tried various methods to isolate and identify viruses. Isolation continues to be a complex research area, with many thorny problems associated with what humans wanted to do versus what nature allows. Viruses, by their nature of being among the smallest living organisms and their abilities to mutate and change over time, almost at will, have confounded the search. Historically speaking, we have made inroads into this problem, but before the IVDS system was invented, little real progress was made in the sampling and counting of viruses.

Many methods have been tried over the years to extract, purify, and concentrate viruses for analysis (Wick et al. 1997b). Most exhibited major shortfalls in selectivity, their applicabilities to all viruses, or their recovery. Of 18 familiar processes, only 2 show high selectivities and excellent recovery characteristics and are universally applicable for all viruses. The two processes are sedimentation rate centrifugation and density gradient centrifugation. The other 16 processes have inherent weaknesses that are fundamental to their processes and thus presented limited utility for virus detection. Some of these limitations may never be resolved.

A successful methodology should make use of processes that have achieved high marks in all three categories (selectivity, applicability, recovery). The IVDS device uses physical methods that earned high marks in all three areas. The physical properties are not dependent upon complex reactions and thus are reliable bases for building a virus detection methodology.

1.2.5 History of IVDS

The IVDS represents a new concept and system for detecting and analyzing viruses. First proposed in 1993, it utilized advances in historically successful technologies and combined them with innovative filtration technologies and computer systems (Wick et al. 1997a; Wick et al. 1997b). The IVDS device builds upon the successful sedimentation rate and density gradient centrifugation technologies dating from the early 1930s and greatly advanced since then. Combining a new ultrafiltration system with well-known differential mobility analyzer (DMA) and condensation nucleus counter (CNC) technologies, it was possible to build a new system for extracting, purifying, and counting viruses.

This new IVDS device was developed in four distinct integrated stages: (1) collection, (2) extraction, (3) concentration, and (4) detection. Because the IVDS system makes use of the fundamental physical properties associated with viruses, much of the risk historically associated with virus detection is minimized. The device utilizes well-known and highly successful commercial equipment and thus presents little or no instrumentation risk. The only risk—and a minor one—is the integration of some of the parts into an advanced prototype. This integration step follows many similar and successful examples and did not at the time appear to present any unusual difficulties.

1.2.5.1 Collection Stage

The first concept considered for the sample collection stage was the Army's XM2/XM19 collector. It was an established system and considered appropriate. Until a new method was identified, it would at least approach the target specifications of the IVDS system. It should be noted that the IVDS can count virus particles for many kinds of liquid samples and from any source. The initial use of the XM2/XM19 avoided the need to develop a new technology at that time and it was already an integral part of the biological integrated detection system (BIDS). It was thought that because the sample volume needed for the IVDS was very small (about 10 µL/minute) the IVDS sample could be split from the sample stream of the XM2 without effect on the other components of the BIDS.

The development and inclusion of the IVDS matched the overall philosophy and design approach of the BIDS although the composition of the BIDS changed constantly. The intent was to field a system of already-developed components and replace the components as improved capabilities become available. If the XM2 had to be replaced, the IVDS would simply adapt to the sample stream provided by a new system.

1.2.5.2   Separation Stage

The separation (extraction) stage employing ultracentrifuge technology starts with the collected material containing viruses. The technology originated in the 1940s at the Oak Ridge National Laboratory (ORNL) when it was employed to separate nuclear fuel materials. Over time, the technology was improved and applied to other uses. Ultracentrifuge technology has advanced in recent years and the application to virus separation represents a renewed use. Patent applications filed in 1996 cover the principal ideas for using this technology for separating viruses.

The ultracentrifugation process for IVDS takes advantage of the virus window (Section 3.2.2 in Chapter 3), a phenomenon by which virus particles are separated from other materials based on specific physical properties. The virus groups are in reality separated into a three-dimensional (3D) "address" that is unique for the various virus families and most likely for the viruses within the families.

In the 1960s, engineers at ORNL perfected a centrifuge technology designed specifically to extract viruses from biological backgrounds. They designated the virus window as an area bounded by density of 1.175 and 1.5gm/mL and sedimentation coefficients of 100 and 10,000 Svedberg units. This window represents the true 3D physical location of virus material within a centrifuge instrument.

Ultracentrifugation offers several advantages. It provides universal capture of all viruses, with proven efficiencies exceeding 95%. This is very important because we were dealing with very small quantities of material. The technology permits recovery approaching 100% at flow rates up to 320 mL/minute. The flow rates used in IVDS were expected to be lower and thus recovery rates approaching 100% were expected to be routine.

Ultracentrifugation also provides a high degree of physical separation of viruses from background components and does not depend upon biochemical reagents. Dependence on reagents complicates methods and introduces potential for error. Reducing reagent use has several immediate advantages, the first of which is eliminating the need for highly skilled technical personnel, thus removing the human error element from the process. A process without reagents is far simpler and requires less training to operate.

Elimination of reagents also improves the accuracy and output rate. It reduces the logistic elements of storing and replacing consumable components in an active environment. This is a very attractive feature considering the expense and monitoring required for reagents that have limited shelf lives.

Members of 20 viral families are considered generally pathogenic to humans. After plotting the 3D addresses of the virus positions in the window, two key conclusions are possible: (1) all of the viruses lie in an area free from interference from all other components or background materials and (2) the viruses are separable, with surprisingly little overlap of the 3D addresses. In other words, knowing the density and size of a detected virus particle pinpoints it to a particular virus family in most cases.

During operation, the centrifuge operates as a flow process. The stream of liquid from the collector stage flows continuously into the rotor assembly surrounding a spinning cylinder that contains a solution of a specified density. As the cylinder of solution rotates, the material it contains assumes a density gradient. The viruses separate according to their 3D addresses. The isolated viruses are sent to the detector in a subsequent stage to be counted. Because they occupy unique locations, they may also be identified. Thus the numbers and identities of the viruses in a sample may be analyzed.

1.2.5.3   Purification Stage

The ability to connect the purification (concentration) stage known as ultrafiltration (UF) with the previous extraction step provides a unique edge for the IVDS. The system allows the exquisite purification of target viruses from an overwhelmingly high background of extraneous material. After ultracentrifugation extraction stage separates the materials of interest by taking advantage of the virus window, a sample is further purified and concentrated via the UF process. UF accomplishes two tasks. It first separates viruses from subvirus-sized particles, then concentrates them into a small volume of liquid. The concentration feature enhances the sensitivity of the IVDS.

Early issues with filtration are discussed in Section 1.2.5. Filtration was a limiting issue until the correct pore size for virus purification was determined and tested. Pore size is critical to the success of this process. The size range for viruses of interest is 22 to 500 nm. These viruses fall within the unique virus window observed during ultracentrifugation.

A sample solution may contain interfering proteins and dissolved salts with apparent sizes up to 20 nm, so pore size control of the UF process (particularly at this point) is critical. The 20 nm particles must be allowed to pass through the filter and the 22 nm particles must be retained. Experimental efforts in the 1990s solved these issues and resulted in a material that satisfies this requirement.

1.2.5.4   Detection Stage

The detection stage involves three components: (1) an electrospray (ES) nozzle, (2) a differential mobility analyzer (DMA), and (3) a condensation nucleus counter (CNC), sometimes called a condensation particles counter (CPC). Detection serves two purposes. First, it counts the individual particles and second, it determines the sizes of the detected particles.

When detection is combined with ultracentrifugation, the viruses are separated according to their unique 3D addresses and the counting device counts a specific group of viruses and allows them to be identified. When used without ultracentrifugation, a detector counts all the virus particles present. Without a specific 3D address developed during ultracentrifugation, no identification is possible without further steps because of the way viruses behave. The combination of ultracentrifuging and detection represents a new way of identifying and counting viruses.

TSI, Inc., based in St. Paul, Minnesota, established itself as a source for the ES, DMA, and CNC components and various integrated combinations. As with the other stages of IVDS operation, using commercial components reduces developmental risks. The ES and DMA capabilities were available commercially in single integrated units. It was also possible to arrange for the integration of all three components. The fully integrated system became the IVDS-Beta and three instruments were made.

This combination of current and reliable instrumentation provided a bold physical method to detect viruses. It is not dependent upon obscure chemical reactions or times required for reagents to interact. The method depends solely on the inherent physical properties of the particles that are unlikely to change and thus allows the particles to be detected and identified. This represents the first real change in virus detection in almost 70 years.

1.2.6   DEVELOPMENTAL RISKS

1.2.6.1   Collection Stage

The XM2 collector was a fully developed device and has been tested for virus collection. The XM2 samples airborne particles in the range of 2 to 10 microns at a collection rate of 1,000 liters/minute. IVDS can work with any new collector or can process a liquid that is directly inserted to count and identify virus particles. The collection stage was considered a low risk to the project because it was fully developed and the integration to the IVDS did not involve any new hardware or processes.

1.2.6.2   Extraction Stage

The work of ORNL in the 1960s demonstrated a 95% or higher virus recovery at flow rates of 100 liters/hour or more in 30 minutes or less. This successful work was not pursued further because of the lack of effective ultrafiltration and calculation technologies and limited need for a virus detector then. Requirements and political and military events in the 1990s changed this position and the advances of the various technologies used in the IVDS made this process a low to moderate risk to the IVDS project because it had not been used in this manner and mechanical connections and testing may have unknown issues that would need to be solved. In practice it was determined that this stage was unnecessary because of the success with ultrafiltration.

1.2.6.3   Purification (Concentration) Stage

The ultrafiltration membranes and techniques available gave the expectation that they would provide exceptional control of pore size and a tangential flow process. This technology had been demonstrated and needed only integration into the IVDS. The integration was not considered difficult. It required precise manufacturing and thus presented a low to moderate risk. It is interesting to consider this early assessment after time passed and various issues and limitations were resolved. The assessment of a moderate risk was appropriate and justified because of the ability of viruses to pass through many of the filter membranes initially considered adequate.

1.2.6.4   Detection Stage

The combination of the ES, DMA, and CNC devices was assembled by the TSI Corporation. The individual technological components were well tested and demonstrated and considered suitable for the IVDS. At the time, detection was considered a low-risk stage because it was fully functional. From a historical view, this stage turned out to be the most rugged and reliable of all. The IVDS-Beta unit was in service for several years until it was replaced by the updated commercial unit that continues in service. The Beta unit was taken out of service only because of the improved features of the commercial unit, not because it was less reliable.

1.2.7 SUMMARY

This summary was written in 1997 and remains true. "Many methods have been proposed and used historically for the separation and identification of viruses. All of these attempts have not been entirely successful and certainly not very helpful to military requirements that call for a quick analysis and assessment of a virus attack. The important issue has not been solved and a simple off-the-shelf solution does not look promising. A bold new concept based upon historically successful processes is in order and is suggested which uses a combination of technologies, integrated into a new system, called the Integrated Virus Detection system (IVDS). This device represents an enormous and fundamental breakthrough in virus detection. The complex issues associated with the historical approaches have been solved and the IVDS device can be expected to be the virus detection and characterization device of the future" (Wick et al. 1997b, p. 18).

1.3 IT WORKS! THE EUREKA MOMENT

I was working on the new IVDS in the lab late one night, trying to generate an MS2 peak. Nothing was working despite my use of many, many variables involving capillaries and solutions and dilutions. I was tempted to shake the device to settle my dispute with it and thankfully ran out of 750 KDa filters. I was not happy and looked for an alternate. I had a good supply of 100 KDa filters and used one. I was annoyed at having run out of the size of filter confirmed by the literature as being the correct one to use. I was watching the graph develop as the instrument followed its now-familiar mechanical routine and saw for the first time a very distinct MS2 virus detection. Well, that was something. I made a few notes, felt great relief that the instrument worked, and went home. I could hardly wait for the next day.

The day after the discovery that MS2 passed through 750 KDa filters was very exciting. Looking around for a sample to test, we decided to test our drinking water. We were never certain what it contained. Also, it was easy to prepare. We started by running a small 20 mL sample through the instrument. The negative results were very discouraging. Outside the lab window was a runoff pond created after a new building was constructed. Grass had grown around the pond and it attracted several ducks. A same-sized sample processed from 20 mL exhibited many viruses.

The next several years involved the testing and expansion of our results to many different viruses, configurations, and characterizations of the IVDS instrument. Patents were submitted. The first patent license was granted and it was clear that the original guidance for a device to detect viruses was met and even exceeded.

During this period, I had an opportunity to give a presentation to a group in San Diego that was looking to support and fund a new technology related to biological defense. They were biased toward new polymerase chain reaction (PCR) and molecular biological methods such as emerging chips. I had just started the presentation when I was interrupted by a loud harrumph and a comment that "this will never work." I was equipped with graphs, charts, and facts, so I proceeded. When I finished, the group was very pleased, and the IVDS was awarded a first place among all the new technologies. The lesson to take home is that the facts will carry the day. Confidence comes from hard facts.

1.4 FLOWCHART FOR VIRUS DETECTION

1.4.1 INTRODUCTION

Discussion of the flowchart for detecting viruses (Figure 1.1) at this point is important for several reasons. First is the depiction of the process from sample collection to electronic reporting of results involving several steps. This flowchart resembles most biological sample processing schemes with the exception of the detailed molecular biology steps that are unnecessary for mechanical detection methods like IVDS.

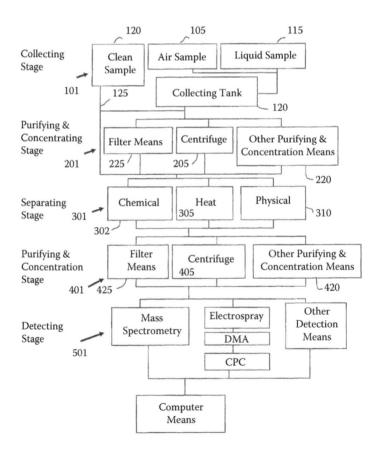

FIGURE 1.1 Sample flowchart for virus analysis.

Each step involves choices based on sample type. In the simplest procedure, the flow-chart is reduced to three or four steps to yield detection and the test may only take a few minutes. More steps are required for complex samples.

The IVDS instrument operates at the detecting stage. Notice that the chart cites other methods in the detecting stage. Some of these other methods may be indicated for detailed detection requirements (Wick 2014). Most of the time, samples are processed through the IVDS to determine whether a virus or viruses are present, then other methods are used for identification. Likewise, each other stage has choices for processing samples and preparing viruses for detection and analysis. Let us take a look at these stages.

1.4.2 COLLECTING STAGE

This stage permits three choices (Figure 1.1). A "clean" sample, one that is free of other material that requires removal, can proceed to the second stage without further processing. Air samples and liquid samples generally contain other material that requires removal or concentration. For example, a virus that is already in a clean liquid requires no further collection steps. Air samples need to be scrubbed of viruses and the virus-like particles must be transferred into a liquid. Because virus loading in open atmosphere is frequently low, the sampling needs to be done over time and the virus-containing liquid should be accumulated in a collecting tank. Likewise, a liquid sample may need processing. A few viruses in a gallon of liquid are not necessarily evenly distributed throughout the liquid, and the whole gallon needs to be reduced to a few milliliters.

1.4.3 FIRST PURIFYING AND CONCENTRATING STAGE

The two main methods for purifying and concentrating viruses are filtering and centrifugation. Other methods are available but they are more difficult and time consuming and not useful for our purposes.

Multistage filtration can remove background materials and successfully isolate a virus, then further purify the virus by filtering the subvirus-sized materials from the solution. This is the preferred step and works well in most applications with minimum effort and time expenditures. Centrifugation takes specialized equipment. A centrifuge may be simple or complex; some samples require ultracentrifuge equipment.

During purification and concentration, a gallon of liquid is reduced to a few milliliters in volume, usually by size differential filtration. This processing of large volumes of liquid has been successful with large water samples. A virus sample in a liquid is reduced in volume (concentrated) and the viruses are separated from the larger particles (purified) and ready for the next stage.

1.4.4 SEPARATING STAGE

This stage is indicated when additional separation of virus from background is required. This separation stage may be redundant in practice, but is included here for completeness. The physical technique is often filtration. In some cases, difficult background materials such as large proteins are removed by enzymes or other means such as heat or chemical treatments.

Because the objective of the analysis is to count the viruses, background material must be removed first. If the objective is to count the nuts in a jar of chunky peanut butter, it is easier to remove the peanut butter first. This can be done easily by placing the jar in hot water, waiting until the peanut butter melts, and pouring the contents through a screen of a size that will retain the peanuts. It is then easy to count the peanuts. The same concept may be applied to many virus backgrounds.

1.4.5 Second Purifying and Concentration Stage

This second purifying and concentration stage is indicated for samples that went through the separation stage. Additional concentration and purification can simplify the analysis later. This second step is frequently not needed; the first purifying and concentration stage is usually sufficient.

1.4.6 Detection Stage

After the earlier stages are completed, the sample is ready for insertion into the IVDS that then examines it and reports any virus or virus-like particles found. The output will include all the virus and virus-like particles in the sample and send the data to a computer for cataloging and further processing. This step is completed in minutes.

Mass spectrometry is indicated for users who require accurate identifications of strains of viruses. This process is detailed in *Identifying Microbes by Mass Spectrometry Proteomics* (Wick 2014). Samples can also be identified by other methods such as multichip, PCR, and molecular probes.

1.4.7 Data Processing

Sample information is conveyed directly from the IVDS unit to a computer and stored. Software can be used to redirect the results to decision makers, use them for further analysis, or archive them.

1.5 CONCEPT OF IVDS OPERATION

IVDS is effective as a first-line detector to determine the presence or absence of viruses. Many other methods can be used later to confirm or identify viruses. When IVDS is used in this manner, the process then provides virus detection and confirmation by two different technologies. IVDS can be used to screen large numbers of samples because it is quick and can detect multiple viruses in a single sample. It can be an important means for detecting unexpected or unknown viruses.

Initially and as described below, the full IVDS operation was more robust. When the ultracentrifugation step was included into the integrated system, it became possible to identify viruses later detected by the GEMMA unit because of the unique 3D addresses based on the physical properties of the viruses. In practice, this step is frequently omitted. The IVDS can be used to detect viruses in a sample and estimate their concentrations quickly. These parameters are usually sufficient for making management decisions.

The presence or absence of a virus is the first question of a first responder or a scientist analyzing a biological sample of interest. The detection of a particular-sized virus is helpful for determining whether to test further. In the case of an infectious particle decisions regarding isolation, quarantine, and other management measures can be made while further tests that take much longer to develop results can be conducted. Frequently, those dealing with infectious agents have no time to wait for results because infections can spread rapidly. Also, governments lack the facilities to isolate and quarantine every individual who may have been exposed to an agent of interest.

1.6 IVDS SYSTEMS AND OPERATIONS

1.6.1 Functions

The IVDS instrument is tasked with the automated detection, identification, and monitoring of submicron-size particles. The present invention performs sampling, detection, and identification of viruses and virus-like agents (such as prions, viral subunits, viral cores of delipidated viruses, and other particles) in bioaerosols and biological and other types of fluids.

1.6.2 Challenges

In the past, it has been extremely difficult to detect and identify viruses without limiting the results to a specific family, genus, and species. Searching for viruses pathogenic to humans in a single environment is also extremely difficult. Additionally, sampling viruses and virus-like agents in air and in fluids increases the complexity of the detection and monitoring functions.

The difficulty of detecting and monitoring a wide range of viruses also varies by environment, and perhaps the most troublesome environment involves field conditions. Detecting and monitoring viruses in a potential biological warfare (BW) threat environment is extremely demanding. Notwithstanding the variations in virulence from virus to virus, a generally accepted figure is that ingestion of 10^4 virons constitutes a significant threat to a soldier who breathes about 1,000 liters (1 m^3/hour) of air. Instruments sensitive enough to allow detection of remote releases of biological agents in a field environment, thereby providing early warning capabilities, allow important safety and operational decisions to be made quickly. Although progress has been made in areas such as optics, the instrument that can fill all these needs remains elusive.

In the past, it has been very difficult to develop and maintains broad-spectrum systems for virus detection that are free from false negatives caused by natural or artificial mutations. The high mutation rates of known viruses and the emergence of new viruses such as Ebola virus must be addressed by any detection method. The potential for deliberate artificial mutations of viruses also exists. Furthermore, virus-like infectious agents such as prions are suspected of causing scrapie, mad-cow disease, and Creutzfeldt–Jakob disease. These prions possess no DNA or RNA, and can withstand 8 MRad of ionizing radiation before losing infectiousness. Other virus-like infectious agents such as satellites possess no proteins. However, detection of all of these varied types of agents must be possible for a device or method to be generally effective for detecting and monitoring virus fragments, prions, satellites, and other particles that are pathogenic to humans.

Detection and monitoring of viruses must also be free from false positives associated with various and diverse backgrounds. For example, a detection system may register background biological debris in a sample as a virus. Analysis of viruses requires a very high degree of purification to overcome background loading and avoid false positives. For example, a BW virus may be buried within other micro-organisms that form biological debris with loadings 10^{10} times larger than the threshold loading for the targeted virus.

Although methods that culture viruses can often be used to increase the virus counts over background, culture methods are too slow for efficient viral BW detection; furthermore, some important viruses cannot be cultured by known methods. In any case, cell culture methods are highly variable and inconsistent.

Viruses may also be extracted from an environment and concentrated to amounts required for detection and monitoring without requiring culture. Generally, when detecting small amounts of viruses in environmental or biological liquids, it is necessary to enrich the concentration of viruses many orders of magnitude (greatly reduce the volume of liquid that is diluting the viruses) and achieve complete removal of non-viral impurities. In the presence of non-viral impurities, even the most sensitive detection methods generally require virus concentrations on the order of 10 femtomoles per microliter or more in the sampled liquid to detect the viruses reliably.

Sampling for airborne viruses is generally accomplished by collecting airborne particles into a liquid using a process such as air scrubbing or eluting particles from filter paper collectors into a liquid. Because collection and subsequent separation and detection methods are strongly affected by the adsorption of viruses to solids in aerosols and by solids association in water, designs for sampling of air for viruses must follow stringent requirements.

In contrast, no special devices or processes are generally needed to collect virus samples in liquids. For example, only a standard clinical hypodermic needle may be needed to sample blood and other body liquids for viruses. For sampling of bodies of water or other accessible liquids, sample collection may not be an issue. Virus extraction is achieved by collection on a filter.

Currently, we have no simple and rapid method or device for detecting pathogenic viruses in a BW environment. Rapid detection translates into protection, more reliable and simplified strategic planning, and validation of other BW countermeasures. Previously known detection methods using biochemical reagents are impractical in the field, even for trained virologists. Additionally, reagent-intensive approaches, such as multiplex PCR, low-stringency nucleic acid hybridization, and polyclonal antibody testing may increase the incidence of false positives several hundred times.

The hypervariability and rapid mutation of viruses and emergence of new and not-yet catalogued viruses further preclude methods such as PCR and immunoassays based on biochemical principles from achieving broad-spectrum detection of all known, unknown, sequenced, and unsequenced viruses. A highly reliable automated system that can handle changing sample conditions and provide rapid response is needed. Computer control of instrumentation, data collection, and interpretation provides several advantages including increased operator safety, simultaneous multiple location detection, decreased training requirements, and improved result consistency.

1.6.3 OBJECTIVES OF INVENTION

Based on the challenges described above, the IVDS was intended to provide universal monitoring and sampling of viruses and virus-like agents. Another objective was to achieve rapid automated detection and identification of viruses based on their physical characteristics, thus eliminating the needs for biochemical reagents and complex assays. We also wanted to develop a system capable of detecting known, unknown, and mutated viruses.

The ability to detect many viruses or virus-like particles simultaneously and thus provide a method to monitor natural virus loading over time was critical. This feature would allow tracking of changes of virus loading over time to determine trends. Changes from a normal or natural situation would be immediately recognizable and allow remedial action if required. These and other objectives have been achieved by the IVDS. It is capable of detecting submicron particles by

- Collecting a sample (this step is trivial for waterborne particles but requires certain steps in cases of airborne particles and complex media)
- Extracting existing submicron particles from collected samples based on density or size via filtration
- Purifying the samples by size through concentration and filtration
- Detecting submicron particles based on their differential mobilities
- Automatically transmitting data to a computer; and if required, disseminating data to multiple sites electronically

The IVDS can count the extracted and purified submicron particles and identify them. It can also act as a sample collector, an extractor for removing submicron particles from collected samples, a purifier that concentrates the particles, and a detector utilizing the 3D addresses of purified particles. The IVDS is a valuable addition to the field of virus detection. Details of its operation in several real-life situations are discussed in later chapters.

1.6.4 OPERATIONAL DETAILS

The IVDS (Wick et al. 2000; Wick et al. 2002a; Wick et al. 2002b; Wick et al. 2007) constitutes a method and device for the detection and monitoring of submicron particles. It can collect, extract, purify, and detect viruses that are pathogenic to humans. The method and device include an automated system for the rapid detection and monitoring of viruses and achieve synergistic use of centrifugation and ultrafiltration virus purification methods.

Centrifugation separates viruses from non-viral impurities by using powerful centrifugal forces to break virus particles away from solids, cells, and cell substructures without highly concentrating samples. Ultrafiltration performs purification and concentrates samples after the viruses are separated from extraneous material by the centrifugation step. These systems separate viruses from background material based on density (via isopycnic banding centrifugation) and size (via ultrafiltration).

Neither density nor size alone is sufficient to distinguish a virus from extraneous particles. For example, the densities of proteins overlap those of viruses. The sizes of microsomes, glycogens, other cell structures, and macromolecules overlap those of viruses, but none of those particles overlap viruses in both density and size. By combining measurements of both properties, a virus window can be constructed. The data from the window can be used to identify viruses in samples. Use of the virus window is discussed in Chapter 3.

One feature of the IVDS is the ability of density-gradient centrifugation (isopycnic banding) separation that allows a single stage to separate viruses from non-viral material based on density and also achieve a substantial degree of protein removal regardless of density. This allows removal of proteins that are ubiquitous in sampled fluids and have densities that overlap the densities of viruses. A substantial removal of proteins at this stage complements the subsequent removal in the ultrafiltration and DMA stages. Figure 1.2 shows the IVDS 100. The system includes a collection stage 101, an extraction stage 102, a purification or concentration stage 103, and a detection stage 104. These four stages are detailed below.

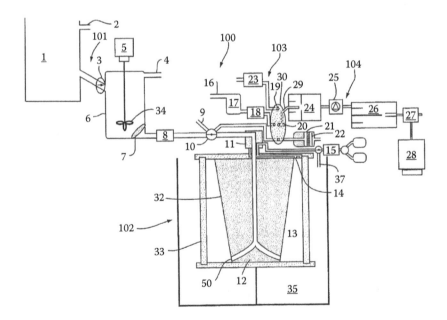

FIGURE 1.2   Integrated Virus Detection System (IVDS) diagram.

1.6.4.1 Collection

In the collection stage 101, a collector 1 is used for aerosol or gaseous fluid sampling. The collector 1 samples airborne particles having sizes that optimally carry viruses—2 to 10 microns at a collection rate of 1,000 liters per minute of air. Collection of the submicron-sized virus particles in the collector is facilitated by the ability of airborne viruses to travel in or on aerosol particles larger than 1 micron. In exceptional cases where a virus is not rafted on a supermicron particle, the danger of transmission by inhalation is generally reduced because of the distribution of submicron particles in the atmosphere and the difficulty of capture by the lungs. The collector has an inlet connected to a water source such as a tap or water purification system. The collector scrubs the collected particles with the incoming water from the inlet. Examples of collectors are the US Army's XM2 and the Spincon made by the Midwest Research Institute.

In many applications other than aerosol sampling, specimens possibly containing viruses may be obtained without need for a formal collection process, e.g., if a specimen is already in liquid form. Examples are blood samples obtained in clinical settings and other body fluids such as mucus, semen, feces, lymph, and saliva. Other liquid samples may be obtained from municipal water supplies, rivers and lakes, beverages, and high-purity water used in microelectronics manufacture. In summary, to detect viruses or other particles in aerosols, another step and additional equipment not required for liquid samples is indicated.

Tubing connects the collector to a holding tank having a blender or homogenizer. The collector's aqueous stream output containing the scrubbed particles is on the order of 1 to 10 mL per minute. It is pumped through the tubing (preferably Teflon or polysiloxane-coated to reduce adsorptive losses) to a 1-liter holding tank. Alternatively, the tubing may lead directly to the extraction stage 102.

The solids in the aqueous stream are broken up by a bladed homogenizer in the holding tank or alternatively by forcing the stream through an orifice. A surfactant or amphiphile is added at the inlet and preferably mixed with water before its entry into the holding tank. The surfactant or amphiphile breaks down the structures in the aqueous stream. The amphiphile should have a low boiling point to allow its easy removal at a later stage. The most preferred amphiphile is diethylene glycol monohexyl ether. A base may be added to increase the pH of the homogenized liquid and help decrease aggregation.

Upon leaving the holding tank, the aqueous stream passes a screen filter that regulates the output of the tank. The preferable screen filter is of 10-micron mesh and made of stainless steel or another corrosion-free material. A pump designed for moving liquids through a tank draws the aqueous stream from the holding tank and moves it through the screen filter.

1.6.4.2 Centrifugation

Beyond the pump, a three-position computer-controlled switch allows the discharge from the pump to flow into a centrifuge rotor (first switch position). To understand the need for the second and third positions of the switch, it is necessary to realize that the gradient imprisoned in the rotor after centrifugation can be divided into

(a) the portion containing one or more ranges of densities within which the particles to be detected are expected to lie that will be sent to the next stage and (2) the remaining material that will be discarded.

Thus, for example, in detecting viruses pathogenic to humans, the useful portion of the sample could be the part of the gradient corresponding to densities of 1.175 to 1.46g/mL as discussed elsewhere. Alternatively, a subset of this range could be useful if only certain viruses are being analyzed.

The second switch position allows the useful part of the gradient to flow onward. The third position allows the discarded portion of the gradient to flow from the rotor through a port. If desired, the port can incorporate means to recycle density gradient materials. With the switch in the first position, the screen-filtered sample from the pump travels past the switch and enters the extraction stage 102 where it enters a liquid-cooled coaxial seal.

After passing the coaxial seal, the aqueous stream enters at the upper shaft of a zonal centrifuge rotor such as Beckman's CF-32 or Z-60. The rotor is inserted into and spun by a centrifuge such as a Beckman Optima XL-100K. For large sample volumes containing small quantities of viruses, for example drinking water sources, the IVDS uses continuous-flow density gradient ultracentrifugation with the Beckman CF-32 rotor. For other applications, ordinary zonal centrifugation with a Z-60 rotor is preferred.

In a special seal and bearing assembly, fluid inlet and outlet streams access an annular space between a core and rotor wall through the coaxial seal assembly and via port. Density gradient solutions, sample liquid, and the displacement fluid are sequentially pumped into the annular space. Density gradient solutions are loaded from the port through the inlet. The pump adds sample liquid. A density gradient solution is any liquid that permits the separation of viruses; examples are sucrose and cesium chloride.

In continuous flow operation, the virus-containing liquid stream is pumped from the collection stage 101 and flows continuously over the density gradient in the rotor. The viruses sediment out of the stream, banding into the density gradient according to buoyant density. This pumping of sample into and from the rotor can be performed while the centrifuge spins at high speed. The continuous stream allows a large volume of fluid to flow through the annular space, permitting virus material to be captured in the gradient, even at small concentrations of viruses. In ordinary zonal operation (not continuous flow), a sample does not flow continuously into the rotor for long periods of loading. Instead, the entire sample volume, which must be less than the annular volume of the rotor, is loaded into and enclosed by the rotor.

The rotor volume is then closed off before acceleration to high speed. In both situations, this is called the loading phase of isopycnic banding separation. After loading and centrifuging to achieve banding, the virus-containing bands are recovered by displacing the bands sequentially, with lowest density bands exiting first and highest density last. The density of each virus or particle uniquely determines its position in the exiting stream and thus the timing of the detection of specific particles provides density information.

1.6.4.3   Density Gradient

A fresh gradient is loaded into the rotor by pumping a low-density fluid containing no cesium chloride into the rotor. Two fluid tanks and a mixing valve (15 in Figure 1.2) mix a high-density fluid typically containing about 60% cesium chloride with a low-density fluid at a variable high:low ratio. A computer control increases the ratio over time until loading is complete. The fluids pass through the fluid entry ports at the top of the annular space. Concurrently, the rotor spins at a low speed (~4,000 rpm). The speed is controlled by the timer control system in tandem with fluid entry and displacement.

After the fresh gradient is loaded, the control system actuates valve that move fluid through the rotor in the opposite direction, pumping sample from the holding tank through the switch (in first position) to the bottom entry port and upward through the annular space, entering at the bottom end and displacing fluid at the top of the rotor out through the discharge port. After establishing flow reversal, the control system initiates and regulates the centrifuge to a preferred rotational speed of ~60,000 rpm for a B-series rotor. In extremely dry environments, water exiting the centrifuge may be recycled back into the system by pumping back into the collector where it can be used for air scrubbing.

At a rotation rate of 60,000 rpm and flow rate as high as 6 liters per hour, over 90% of all virus enters the gradient from the sample fluid stream and remains imprisoned there. After 10 to 30 minutes of operation allowing as much as 3 liters of sample fluid to pass through the rotor, the inflow and effluent flows are shut off and the high-speed rotation continues for an additional 30 minutes to band the viruses in the gradient. The centrifuge controls are actuated by a timer-regulated control system, usually a standard PC interface.

In operation, sample liquid is introduced into the density gradient within the centrifuge rotor at the low-density end of the gradient, and each particle or molecule penetrates the gradient at a rate that increases with the mass of the particle and its density. In the case of a protein molecule, the mass is much smaller than that of a virus by at least an order of magnitude, and the density is about the same as that of a relatively low-density virus. Accordingly, the rate of banding for proteins is much slower than for viruses.

The centrifugation continues just long enough for the smallest virus particles of interest to have enough time to band to the desired resolution in the gradient. This is typically within 1 to 5% of the equilibrium position. The proteins will then primarily be on the low-density side of their equilibrium position because they started on that side. Because the equilibrium positions of most proteins in a gradient are nominally about 1.3g/mL, at the end of this shortened operating time most proteins are positioned considerably lower than 1.3g/mL. The proteins are at positions where they are not collected and sent to the next stage as they are outside the virus window. Accordingly, the density gradient centrifugation step takes on some of the properties of a combined two-stage density-gradient sedimentation coefficient separation.

After the viruses are banded, the centrifuge is decelerated to low speed, and the gradient is recovered by pumping the dense fluid (preferably 60% CsCl) from the gradient supply system to the outer edge of the annular space. The dense fluid displaces the gradient, with low-density bands exiting first followed by high-density bands. After gradient removal, the high-density material in the rotor is displaced by low-density fluid entering from the inner rim of the annular space that displaces the high-density material from the outer edge of the annular space. The procedure is complete in a few minutes, and the cycle begins again with the loading of the density gradient at low speed.

1.6.4.4   Virus Window

Ultracentrifugation achieves the desorption of viruses from fomites (materials that may be contaminated) and universal capture of all catalogued and uncatalogued viruses with capture efficiencies exceeding 95%. Ultracentrifugation is not dependent on biochemical reagents and provides a high degree of virus separation from background components. It also generates virus density information by determining the y coordinate of the virus window plot. A commercially available coaxial seal is Beckman's zone assembly (Part 334241).

The results of the extraction of viruses from biological background by ultracentrifugation are analyzed by means of a virus window defined as a density–size (p-d) or density–sedimentation coefficient (p-S) plot of biological components. The x axis shows either size d or sedimentation coefficient S; the y axis denotes density p; see Chapter 3, Figure 3.3. Mammalian virus densities are between ~1.175 and 1.46g/mL density and have diameters between ~22 and 800 nm; expressed alternatively, in this size range, the sedimentation coefficient is between 120 and 6,000 Svedberg units.

The virus window is a very useful tool because shows how viruses can be separated from non-viral backgrounds and also shows how different virus families are substantially separable from each other. Each virus family within a virus window is distinguished by a specific rectangle with minimal overlapping among the 20 family rectangles. Accordingly, if density and size are known, a detected virus particle may be pinpointed to its family in the virus window.

In any case, particles whose densities and sizes both fall within virus window ranges can be presumed with a high level of confidence to be viruses. When counts are registered in the detector of the IVDS after centrifugation for density in a range of ~1.175 to 1.46g/mL and further selected by the differential mobility analyzer (DMA) for sizes between ~22 and 800 nm, we can conclude with a high degree of confidence that these values indicate the presence of viruses in a sample.

This confidence level may be increased further if the density and size fall into a particular region of the virus window known to correspond to a virus. Similarly, other particles of potential interest such as prions, other virus-like particles, and other natural or artificial particles, colloids, cell structures, and macromolecules will frequently occupy unique positions on the density–size plot that may allow them to be separated from other components and thus be detected by the IVDS.

Although only virus to a very large degree fall within the virus window, other background components fall close to the window. These components are microsomes and similar subcellular structures. These components can be eliminated effectively by adding a nonionic surfactant such as diethylene glycol mono-hexyl ether to the collection stage 101 exit stream at inlet 4 of Figure 1.2. The surfactant solubilizes the microsomes and membrane fragments.

Because the recovery of viable viruses is not necessary, release agents can be used. These agents are preferably organic solvents and surfactants, usually low-molecular weight amphiphiles such as diethylene glycol mono-hexyl ether. The release agents provide several useful effects. First, they break up and even dissolve cellular substructures such as microsomes, ribosomes, lysosomes, peroxisomes, and mitochondria whose sizes and densities are similar to those of viruses and set the limit on the resolution required to detect viruses.

Second, upon dissolution of the lipid envelope with such agents, the increase in virus density is significant (density of the viral core—the virus minus its lipid envelope—is generally significantly higher than that of the enveloped virus). In the case of the Hepadnaviridae, for example, the density may be ~1.25 to 1.36. Both effects serve to further differentiate viruses from microsomes in the virus window plot, first by acting to eliminate the microsomes and second by increasing the difference in density between viruses and background microsomes.

Third, release agents enhance the desorption of viruses from solid matter; this is particularly important in the detection of airborne viruses. Release agents can also break up aggregates of viruses, especially encapsulated viruses. The IVDS minimizes this aggregation problem without the need for release agents. The centrifugation can be performed without pelleting. Consequently, buoyant density and thus isopycnic banding are not greatly affected by aggregation under these circumstances. Indeed, banding times are favorably reduced in the case of aggregation, and techniques can be applied to take advantage of the reduction within the broad context of the IVDS.

Any aggregation generally produces only a small shift in and/or broadening of the resulting virus bands. The portion of this exiting stream that contains the virus window is pumped to the purification stage 103. The position of a particle along the stream indicates its density. The useful part of the stream during general virus detection where the range of 1.175 to 1.46 is passed to the next stage is ~10 mL; thus, this stage does not produce a large increase in virus concentration although it does effect a very large increase in the concentration of viruses relative to other non-viral components.

Although feasible, an additional centrifugation to separate particles by sedimentation coefficient for the virus window x coordinate value is not necessary. The differential mobility analyzer (DMA) described below provides rapid analysis of particle size. Additionally, separation of viruses from soluble proteins can also be performed in the purification stage. An even further separation of proteins and other macromolecules smaller than viruses from viruses can also be done by tuning the supersaturation in the CPC so that it will not detect macromolecules as small as proteins. The centrifuge dimension and rotor speed for optimal centrifugation can be calculated. Optimal time is 30 minutes or less and resolution is preferably 0.02 density units (0.02g/mL) or better.

1.6.4.5 Purification

The sample fluid passes from the extraction stage 102 to the purification stage 103. Typically, this could be in the form of 15 pulses, each on the order of 1 to 10 mL in volume and corresponding to a density slice with a width on the order of 0.02g/mL. An ultrafiltration (UF) membrane separates the viruses from soluble proteins (removing the need for a second sedimentation rate centrifugation in the previous extraction stage) and concentrates particles with sizes greater than the pore size into a very small volume of liquid. Also at this stage, the concentrations of soluble salts including those from the sample and the density gradient material (e.g., cesium chloride) are greatly reduced.

The UF membrane may be a Millipore Viresolve or Amicon P or preferably a Pall Filtron Omega with a 1,00,000 MWCO. The water permeability of the UF membrane is on the order of 0.01 mL/cm^2/second/psi so that a membrane area of 0.1 cm^2 yields a flux of 6 mL/minute at 100 psig transmembrane pressure. The UF membrane is incorporated into a housing designed to allow flow rates of 0.1- to 20 mL/min during ultrafiltration.

This results in loading of the filter with particles larger than ~15 nm (including all virus particles), after which the particles are confined within a small front face–side collection volume. A small-volume filtration filter holder 21 such as Schleicher & Schuell's Selectron holds the UF membrane 22. Even better is a filter holder with a design similar to the Selectron but made of an alternative material that does not degrade electrolytically under high voltage.

A four-way positioner 30 in the purification stage 103 allows automated processing of particles in the UF membrane 22. The positioner is driven by a computer-controlled motor that positions the filter holder in one of four ports. In the first position, the UF membrane 22 is positioned to accept the sample flow output from the extraction stage 102. Each 0.02g/mL density slice from the output of the extraction stage is, after passing through switch 10 in the second position, loaded through the UF membrane in less than 2 minutes. Alternatively, larger density slices can be ultrafiltered, requiring appropriately longer times. A standard 0.2 micron pore size filter (such as the Corning Costar) may be incorporated in the connection between the output from 102 and the input to 103 to remove any remaining particles larger than ~800 nm.

When the positioner 30 is switched to the second position, a valve closes off the sample flow. CsCl-free water from pump 18 in tank 17 is passed across the UF membrane 22 using ~5 mL of water with a flux time around 1 minute. This reduces the 30% CsCl aqueous solution surrounding the particles to less than 100 ppm of CsCl and allows recovery of the CsCl for recycling. Additionally, the amphiphile, viscosity additives, and buffer components are reduced in the UF membrane. Ammonium acetate solution of 20 mM concentration in water is preferable for this operation to prepare the liquid for downstream detector stage operation.

On switching the positioner 30 to the third position 19, the pure water (or ammonium acetate solution) is shut off, and a final ultrafiltration is performed to reduce the volume of liquid on the retentate side of the membrane, greatly increasing the concentration of viruses and reducing the volume of liquid to the small quantities required for detection stage. The filtrate in this step passes out through port 23. More precisely, the UF stage 103 is integrated with the electrospray assembly 24 of the detection stage 104 by a punctured disk fitting.

The fitting has a 150-micron hole drilled through a tubular stub in its center. When the positioner in the third position, this hole allows the filtrate to pass out through port 23. When positioner 30 is in the fourth position, the inlet end of the electrospray capillary 29 (opposite the spray tip) is inserted into this 150-micron hole. This fits in a piston-like manner into the stainless steel cylinder of the Selectron or similar filter holder. The cylinder slides over the steel disk and is positioned with a gap of about 100 microns between the steel disk and the ultrafilter surface

In the fourth position 20, in accordance with the above, the UF membrane 22 is positioned for entry of the virus-containing retentate into an electrospray capillary 29 of the detection stage 104. Alternatively, instead of fluid passing directly from the purification stage 103 to the electrospray, an intermediate component may be used to accomplish further purification and/or concentration. A platinum wire may be run from the voltage source of the electrospray unit to the interior of the liquid inside the volume on the retentate side of the UF membrane to establish a current return for the electrospray operation.

1.6.4.6   Detection

The detection stage 104 accomplishes final purification, individual virus particle counts, and determination of particle sizes. The three major components of the detection stage are an electrospray assembly (ES) 24, a differential mobility analyzer (DMA) 26, and a condensation nucleus counter (CNC) 27, also known as a condensation particles counter (CPC). The detection stage 104 can conduct measurements concurrently as the collector processes the next collection cycle.

Passing from the purification stage 103, the retentate enters the detection stage 104 at the inlet of the capillary 29 of the electrospray assembly 24 in the fourth position of the positioner 30. Entry into the electrospray capillary is achieved without passing the retentate through piping that might cause sample losses. The electrospray capillary is ~25 cm in length and the inlet of the electrospray capillary 29 is positioned to the small front face–side collection volume of the UF membrane 22, as described above. The electrospray capillary is then positioned to sample liquid from the retentate side of the filter and the sample liquid enters the electrospray assembly 24.

In the electrospray assembly 24, the liquid sample solution is passed into an orifice or jet of 50-micron diameter. Droplets are ejected under the influence of an electric field. The droplets are typically 0.1 to 0.3 microns in size and their distribution is fairly narrow. At a droplet size of 0.3 micron, the sampling rate is 50 nanoliters per minute, allowing the electrospray assembly to spray the collection volume at a rate around 20 minutes per microliter.

From the electrospray assembly 24, the sample passes to a charge neutralizer 25. The charge on the droplets is then rapidly recovered using an ionizing atmosphere to prevent Rayleigh disintegration. The neutralized ES droplets are then dried in flight, leaving the target virus molecules and/or a dried residue of soluble impurities. From the charge neutralizer 25, the target virus molecules and/or dried residue enter the DMA. The DMA uses electrophoretic mobilities of aerosol particles to classify the particles by size, using the inverse relationship of the mobility of a particle to its size.

In the DMA, particles are carried by an air stream at a set velocity through an electric field created by a charged rod. If a particle is singly and positively charged, it experiences an electrostatic attraction to the rod, which competes with the inertial force of the flow. When the electrophoretic mobility falls in a certain range, the particles pass through a narrow exit port at the end of the charged rod. The particle size range, generally 0.01 to 1 micron, is divided into 147 size channels.

The entire range is automatically scanned in 1 to 10 minutes, generally in 3 minutes. The DMA presents only a possible 3% instrumental error for virus size determination. Additionally, there is a possible size increase due to the covering of virus particles with impurity residue. At an impurity level of 100 ppm, a typical 40 nm virus has a possible error up to ~2% in effective size. If the impurity levels are less than 20 ppm, the error becomes smaller than 1%.

When the primary droplets from the electrospray assembly are 0.3 micron, a 1-ppm soluble impurity creates a 3 nm residue particle. A 125 ppm soluble impurity creates a 15 nm particle. Particles that are 15 nm in diameter can be separated in the DMA from viruses that are at least 20 nm in diameter. Accordingly, soluble impurities must be reduced to less than 100 ppm (0.01%) to avoid background interference with virus signals.

Detection of proteins at levels of 10^{11} to 10^{12} molecules/mL indicates that a sensitivity level for viruses of 10^{10} particles/mL (and possibly 10^9 particles/mL) can be achieved particularly by combining the DMA selection with an adjustment of the Kelvin radius of approximately 10 nm. Impurities of 1 ppm yield a 3 nm residue particle that can overlap protein sizes. Impurity levels of 100 ppm or less are acceptable in the detection of viruses because viruses are several times larger than proteins. Projected sensitivities of 10^{10} molecules/mL (and possibly 10^9 molecules/mL) are based on documented results using proteins. In one of the examples described below, detection of 10^{12} plaque-forming units per milliliter (pfu/mL) was accomplished easily even after dilution demonstrating a detection at a level of 10^{10} pfu/mL.

1.6.4.7   DMA Function

The DMA validates against false positives by changing the dilution and seeing whether particle size also changes. Additionally, the DMA can be used to provide another layer of protection against interference from impurities up to the 100-ppm level. The level of 10^{10} molecules/mL corresponds to 2×10^7 viruses in a 2-mL collection volume of the purification stage 103 and 10^9 molecules/mL corresponds to 2×10^6 viruses. At a collection volume of 10^7 viruses by the DMA or 20 minutes of XM2 sampling, 20,000 liters or 20 m³ of air are sampled. Accordingly, the sensitivity of the IVDS is on the order of 500 viruses per liter of air.

With impurity levels of 100 ppm or less, virus size can be determined by the DMA within about 4%. The detection stage including DMA size determination requires 5 to 40 minutes and may be performed concurrently with centrifugation of samples for a subsequent cycle.

From the DMA, the sample enters the CNC (or CPC) that creates a nucleation effect. The aerosol sample enters and passes through a heated conduit whose atmosphere is saturated in butanol. The sample is routed into a cooled condenser where butanol vapor condenses onto the sample particles that act as nuclei. The saturation is regulated so that no condensation occurs on the nuclei below a critical size; this limits false background counts to fewer than 0.01 particle/mL.

With nucleating particles, condensed droplets grow to micron size and are optically detected using a 780 nm laser diode with a photodetector. If the level of impurities is low enough that the residue particles are below the threshold of detection by the CNC and/or are separated from the target molecules by size, only the target molecules will be registered by the CNC. As the nucleation of droplets does not depend on the surface characteristics of the particles, butanol saturation can be adjusted for a critical size of 0.01 micron radius to minimize background counts from proteins and other soluble impurities.

Response times for step changes in concentration are under 20 seconds, and all components operate in a temperature range of 10 to 38°C. Supersaturation tuning for a 10 nm Kelvin radius threshold in the CNC can be used to cancel the detection of non-viral impurities including proteins if concentration is below ~100 ppm.

The ultrafiltration stage has an output volumetric rate that is very well suited for input into the ES-DMA-CNC particle counter and addresses the strict requirements and narrow range of operating parameters for the ES-DMACNC unit. In recognizing the high value of the molecule-counting and molecule-sizing ES-DMA-CNC unit, ultrafiltration provides excellent samples for the purification and concentration stages that precede detection. The ES-DMA-CNC combination allows particles to be sized and permits improved sensitivity by an order of magnitude over a DMA-CNC combination. Protein concentrations of 10 mg/mL (10^{11} to 10^{12} molecules/mL) can be detected and sized.

1.6.4.8 Automation

The system is controlled by a computer. When data collection and instrument control are handled by a single computer, the computer may vary its mode of operation in response to virus detection. Initially, before viruses are detected, the system passes the entire 300 mL of density gradient from the extraction stage 102 through the UF membrane 22 to scan all virus sizes from 22 to 400 nm. Alternatively, the DMA is bypassed entirely if non-viral concentrations are low enough that tuning of the Kelvin radius in the CNC is sufficient to reduce background.

After viruses are detected, the DMA indicates the sizes. The computer can then trigger the output of the extraction stage 102 to be sampled piecewise in the purification stage 103. By breaking the range of virus densities (~0.3g/mL) into 10 or 15 slices, the density of the detected virus is within ~ 0.02 to 0.03gm/mL—sufficient to narrow most viruses down to a single family.

Following this, the region in the centrifuge output stream surrounding this density can be divided still finer to provide better viral density accuracy. Through database comparison, the system identifies viral families from the measured densities and sizes, and provides output data for detected viruses by density, size, concentration, apparent changes in concentration over time, and if desired audible and/or visual alarms in the presence of detected viruses.

Because the IVDS is automated, it can run continuously for long periods without operator attention. In addition to making continuous virus monitoring possible at a large number of sites simultaneously without the need for scores of virologists, the automation afforded by the IVDS also limits the risks of viral infection of technicians.

Other physical means of separating viruses and other particles from background and/or enriching their concentration that are substantially equivalent in their purification and concentration effects may be applied in addition to or in substitution without departing from the spirit or scope of the IVDS. The means include capillary electrophoresis (purification and concentration enrichment), sedimentation rate centrifugation (primarily purification), hydroextraction (mainly concentration), dialysis (purification and concentration), organic and inorganic flocculation (purification and concentration), and various types of capillary chromatography such as size-exclusion, hydrophobic interaction, and ion-exchange (purification and concentration).

1.6.5   EARLY VIRUS DETECTION EXAMPLES

Analysis of two blind samples was performed, using the ultrafiltration (UF) stage, electrospray (ES), differential mobility analyzer (DMA), condensation particles counter (CPC) triage or the gas phase electrophoretic mobility molecular analyzer (GEMMA) (Wick et al. 1999d).

1.6.5.1   Filtration

Two samples ~1.2 mL in volume and labeled AF0001 and AF0682 were obtained. AF0682 contained viruses; AF0001 was a blank or control, although the virus status of the two samples was not known before testing. All but ~200 µL of each sample was taken and prefiltered through a 0.2-micron Millipore syringe filter with low dead volume. Approximately 400 µL was removed from each sample and processed in an ultrafiltration unit designed and built for this purpose. A 500,000 MWCO was selected to separate and retain the viruses and pass proteins and soluble salts out in the filtrate to be discarded.

Successive diafiltration was used, with each filtration step concentrating retained material into a volume of about 5 µL on the retentate side of the membrane. Between successive filtrations, 20 mM ammonium acetate solution was used to restore the volume to about 400 µL. This strength of ammonium acetate was used for the proper operation of the electrospray (ES) in the detection stage.

After two successive diafiltration steps, two 100-µL portions of the final ultrafiltered sample were collected in two ways. The first 100 µL was obtained by forcing out 100 µL of the retentate volume back through a port forward or upstream of the membrane surface during the final leg of the last diafiltration. This was done after allowing 20 minutes for diffusion of the viruses away from the membrane surface.

The second 100-µL was obtained by using a gas-tight syringe press fit into the filtrate outlet port to push ammonium acetate buffer backward across the membrane from the filtrate side to the retentate side to elute the virus from the membrane. The design of the ultrafiltration unit was such that the retentate side of the membrane remained in a water environment, avoiding an air purge or vacuum flush of the system that could cause irreversible adsorption and breaking of virions.

1.6.5.2 Detection

The filtration yielded three 200-μL samples for each of the two specimens; one before filtering, one prefiltered, and one prefiltered and ultrafiltered. The samples were placed in Eppendorf vials. Each sample contained ammonium acetate to maintain the conductivity necessary for electrospray operation. Because ammonium acetate is volatile, it decomposes and evaporates into ammonia and acetic acid during the in-flight drying of the electrospray-generated aerosol and so does not contribute to the final scan.

Additionally, each sample contained the viral particles and fragments of interest along with non-volatile soluble salts that would yield residue particles after in-flight drying. For example, with the electrospray set at 300 nm droplets, a salt level of 1 ppm would yield an average residue droplet diameter of 3 nm. Very high salt concentrations can increase conductivity and destabilize the electrospray although the data indicate that was not a problem. However, in scans of samples that were not ultrafiltered, the salt peaks from the residue particles extended as high as 15 nm.

Samples were input by inserting a capillary into the Eppendorf vial. Flushes of pure 20 mM ammonium acetate were run before and between samples. Electrospray stability is indicated generally by a current between 200 and 400 nA. DMA-CNC data were compared for typical prefiltered samples before and after ultrafiltration. In both cases, the peaks centered around 9.0 nm; at about 4.2 nm, background peaks due to soluble impurities (mainly salts) appeared before and after ultrafiltration.

1.6.5.3 Results

The results indicate that the ultrafiltration retained the material well, at least as the diameter increased to above 20 nm, and fairly well between 15 and 20 nm. The MWCO curve of the UF membrane rose from 0 to 90+% retention between 10 and 20 nm—a significant finding because intact virus particles in families known to be infectious to mammals are always larger than 20 nm.

This process also produced greatly reduced salt content (expressed as the cube of the diameter ratio) and protein concentrations. The protein concentrations ranged from 7 to 15 nm—the area in which many proteins, particularly those heavier than 80,000 MW, lie. The data under the solid curve are for the non-ultrafiltered sample so the reduction is even more dramatic than it appears in the plot. The 15+ nm fraction was preserved by the ultrafiltration, even quantitatively above 20 nm. The reduced intensity and diameter of the salt peak were also extremely good, and pushed it away from the region of interest.

These results indicate that the ultrafiltration methodology is very effective for removing soluble salts and proteins, while preserving 20 nm and greater fractions. The ultrafiltration reduced the background caused by salt residue particles both in magnitude and location, shifting the peak position for the background from ~11 to ~5 nm, putting it well below the range of interest for virus detection.

The results from environmental source samples were analyzed with the UF-ES-DMA-CNC combination. As a blinded experiment, the components of the two samples were not known beyond the fact that one was a blank containing no viruses and the other was doped with virus or collected from environmental air. Upon analysis of the DMA-CNC data, viruses were detected and counted in Sample AF0682; no viruses were detected in the second sample.

GEMMA data for two ultrafiltered samples show evidence of particles in the 22 to 40 nm range for intact virus particles in Sample AF0682. The counts for them were quite low, translated into 10^{10} particles/mL or 10 femtomoles/mL after filtering. Sample AF0001 revealed essentially no activity in the range of 22 to 40 nm, where background counts are typically 0 or at most 8.

With isopycnic banding information from the centrifuge stage, a double check to distinguish simple protein aggregates or polysaccharides with gravimetric densities of ~1.3 or less from non-enveloped viruses of ~1.4 is possible. By using the full system including the centrifuge, the locations of the viral families in density size space could be mapped systematically, providing a lookup table useful for distinguishing viral and non-viral materials.

After ultrafiltration of 0.5 mL of the known sample from the blinded experiment without prefiltering, a 50-µL sample was analyzed by the ES-DMA-CNC combination. The virus was easily detected and the virions were counted and sized. The size determined by the DMA was 26 nm, in agreement with the literature on scanning electron microscopy (SEM) analysis of Leviviridae. The line width (full width at half max, FWHM) was small—only 2 nm. This indicates that size can be determined very accurately and that viruses of a single type can be distinguished from other viruses and non-viral particles with high reliability.

The ultrafiltered known sample was diluted successively by factors of two and analyzed by the ES-DMA-CNC. The MS2 peak was still distinguished easily, giving a signal-to-noise ratio of ~10:1. The peak amplitudes and areas were plotted as a function of dilution (relative sample concentration with the undiluted sample valued at unity) and demonstrated the linearity of the detection method.

The IVDS-Beta as described has been improved continually the enhancements are discussed in the following chapters. In summary, the system can

- Detect the presence of submicron-sized particles.
- Extract submicron particles from a samples based on density.
- Purify samples by concentrating them based on size.
- Detect virus and virus-like particles via an electrospray assembly and a differential mobility analyzer (DMA).
- Detect and analyze particles of sizes from ~0.01 to about 1.0 micron.

1.7   ORGANIZATION OF THIS BOOK

The chapters follow the history of the IVDS instrument from initial construction through applications; new uses continue to be discovered. Many US patents (Wick 2007; Wick 2010a; Wick 2011; Wick 2012a; Wick 2012b; Wick 2013a; Wick 2013b) have been issued on the original invention and subsequent modifications. The improvements are the subjects of several chapters as they represent milestones in the techniques of virus detection.

The three embodiments of IVDS are described in Chapter 2. The IVDS-Beta was an early instrument we used to work out many practical issues. It proved to be a rugged and useful instrument. The IVDS-Com was the first commercial version and it is still in use today. This instrument featured many improvements over the IVDS-Beta—better sensitivity, sample handling, and software. The future RapidDx-IVDS is portable, more sensitive, and automated, having benefited from more than 10 years of success of the IVDS-Com. The RapidDx-IVDS is more solid state and contains all the relevant technology advances developed in the past 20 years.

Chapter 3 discusses results of the early testing of the IVDS-Beta. The detection of viruses from multiple sources is reviewed along with details of the components and early analysis results. Other methods for characterizing viruses were compared with IVDS and the usefulness of the instrument is demonstrated through examples.

Chapter 4 details the concentration of viruses for counting by the IVDS. A device that could achieve high levels of concentration for samples with low virus levels was invented because of the need to detect viruses to levels of 10^2 or 10^3 pfu/mL levels or lower.

In Chapter 5, we report on the use of IVDS to investigate effects of different treatments such as changes in pH and temperature on the behaviors of viruses. It describes dose-response curves that serve as templates for testing antiviral and contamination treatments.

One of the more interesting uses of IVDS is presented in Chapter 6. A method for isolating a single virus from a mixture of viruses was invented. The intent was to isolate single particles of interest for further study in the areas of vaccine development, specialized detector methods, and data archiving.

Chapter 7 details viruses of various sizes detected and characterized by IVDS. Other nanoparticles identified and characterized by IVDS are discussed in Chapter 8. Chapter 9 explains a technique by which IVDS may detect. Recall that bacteria are a size magnitude larger than viruses and generally not detected and counted using methods such as IVDS.

A major commercial application of IVDS involves the monitoring and characterization of viruses in honeybees. Chapter 10 presents some of the results and explains how IVDS can be used to monitor environmental viruses. The ability to monitor the virus loading and measure the data over time generates a temporal picture of normal activity in an environment and how changes affect virus numbers. Monitoring virus loading over time can reveal the effects of treatment.

Chapter 11 discusses other applications of IVDS and Chapter 12 discusses possible future use of this technology.

2 Instrumentation

2.1 INTRODUCTION

Presently there are three versions of the IVDS instrument. The first instrument will be referred to as IVDS-Beta, the commercial unit will be called IVDS-Com, and a new developing unit is designated the RapidDx-IVDS. IVDS is the common name for all the detectors and the different versions are identified as appropriate.

2.2 DESCRIPTIONS OF IVDS COMPONENTS

The IVDS-Beta is shown in Figure 2.1 and the IVDS-Com in Figure 2.2. The principal concept of IVDS along with its operation and various components are described here. With only minor changes, the concept has remained the same from the first Beta unit to the proposed new RapidDx-IVDS. The IVDS is a modular system built around three principal components:

- Purifier–concentrator
- Electrospray injection
- A size analyzer comprised of a differential mobility analyzer (DMA) to count particles and determine their concentration (initially a condensation particles device was incorporated and recently replaced in the RapidDx-IVDS by a solid state CCD direct counting device)

The IVDS is designed as a modular system to allow upgrades of components as they are developed and improved.

2.2.1 PURIFIER–CONCENTRATOR

The ultrafiltration (UF) system is shown in Figure 2.3. As its name implies, its purpose is to remove extraneous material from a sample and concentrate the virus-like particles to a point where they can be detected. The UF system consists of a fiber-based tangential flow filtration system that reduces the coarser contaminations and reduces sample volume to less than 1 mL. The initial sample volume can range from few milliliters to several liters. Figure 2.4 is a diagram of the UF subsystem.

The most important component is a bundle of hollow fibers whose walls are made of a permeable membrane with wide range of pore sizes. An individual fiber is shown in Figure 2.5. A sample is pumped through the semipermeable fiber and the filtrate is forced through the fiber walls by pressure differentials that may be used in tandem to produce further results (Figure 2.6). The sweeping action of the sample

FIGURE 2.1 Integrated Virus Detection System (IVDS)-Beta unit. This was the first functional instrument combining an electrospray nozzle, differential mobility analyzer, and condensation particle counter.

stream prevents clogging of the fibers. Fibers with pore sizes from 4 μm to 100 KDa diameters are available.

2.2.2 Electrospray (ES) Injection

The sample solution is stored in a cone-shaped vial enclosed in a cylindrical pressure chamber. The chamber accommodates the inlet capillary and a platinum high voltage wire, both of which are immersed in the solution. Maintaining a differential pressure causes the solution to be pushed through the capillary.

The fluid containing the particles exits the capillary and is sprayed through a strong electric field that causes it to form a cone and break into small charged droplets. To prevent corona discharge, the cone is surrounded by CO_2. As the droplets evaporate and dry, they form a plume of particles consisting of whatever virus-like particles (or other large molecules) were contained in the original sample along with solute salt particles. If the concentration of the virus-like particles is not too high, each particle will contain a single virus.

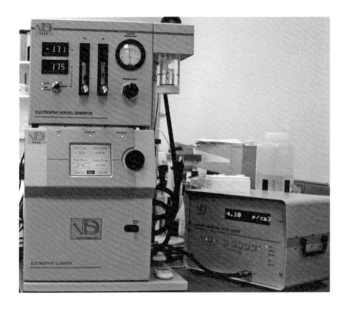

FIGURE 2.2 The second instrument was the first to be commercialized. It incorporated many improvements over the instrument shown in Figure 2.1. The individual components were separated to allow changes as improvements were made.

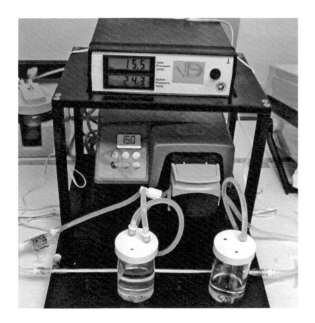

FIGURE 2.3 Ultrafiltration system that utilizes a hollow fiber filter and both isolates and purifies viruses.

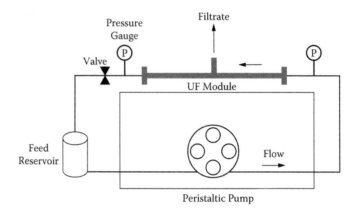

FIGURE 2.4   Figure 2.3 subsystem.

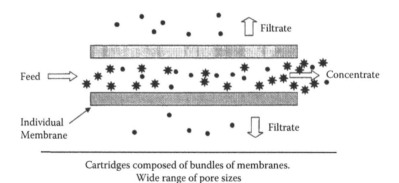

Cartridges composed of bundles of membranes.
Wide range of pore sizes

FIGURE 2.5   Hollow fiber filter operation.

2.2.3   DIFFERENTIAL MOBILITY ANALYZER (DMA)

A differential mobility analyzer separates particles according to their mobility in an electric field. The electrical mobility of a particle is a function of its size and the number of charges it contains. For particles in the size range of few nanometers, the electric charge that a particle will acquire is limited to a single elemental charge. Thus the mobility of the charged particles in an electric field will be limited to their size only. The plumes of the polydispersed particles enter the DMA. The DMA consists of a cylinder with a central rod. A controlled DC potential differential (0–10,000 VDC) is applied between the cylinder and the central rod.

By controlling this potential, only particles with very narrow electrical mobility (size) are allowed to enter the opening slit at the bottom of the cylinder and then enter the particle counter. Figure 2.7 shows the output of the IVDS when injected with the mixture of MS2 virus (23.3 nm diameter) and calibrated by 70 nm diameter polystyrene beads. Figure 2.8 illustrates the resolution capabilities of the IVDS. A mixture

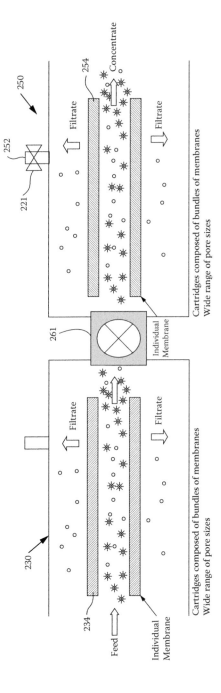

FIGURE 2.6 Dual hollow fiber filter operation.

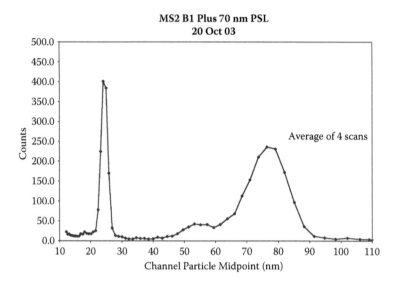

FIGURE 2.7   Output of IVDS injected with mixture of MS2 virus (23.3 nm diameter) and calibrated by 70 nm diameter polystyrene beads.

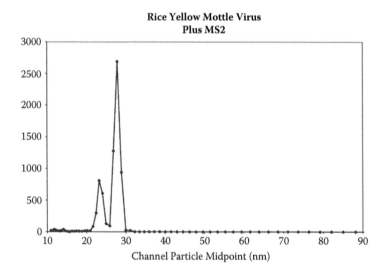

FIGURE 2.8   Resolution capabilities of IVDS.

of rice yellow mottle virus and MS2 was prepared in the laboratory and analyzed. The peaks of these two viruses are clearly separated.

2.2.4   CONDENSATION PARTICLES COUNTER (CPC)

The particle counter is based on the concept that a vapor requires a high degree of supersaturation to condense and form droplets in the absence of particles, Water vapors, for example, must reach supersaturation of 800% before spontaneous (also known as homogeneous) nucleation occurs. In the presence of small particles that serve as nucleation (condensation) centers, condensation occurs at a lower super-saturating level, which is a function of particle size. It can be shown that all of the nucleating particles grow to droplets of identical size that depends on the availability of vapors. The number of particles then is identical to the number of droplets formed. This number can be deduced by measuring the opacity (light transmission) in the volume containing the particles.

In the IVDS, the condensing vapor started as n-butanol (later changed to a water-based system). The RapidDx-IVDS replaces the CPC altogether with a solid state CCD. Nevertheless, in the early IVDS models, the aerosol-laden air from the DMA was passed through a heated chamber (39°C) over a pool of liquid butanol. The butanol-saturated air passed into another chamber held at lower temperature (10°C) where it became supersaturated. The butanol vapors condensed on the particles present to form a "fog" whose density of was a measure of the number of particles present and was determined by measuring the attenuation of light. The supersaturation that can be achieved, and hence the smallest particle that can be activated, are functions of the temperature differential between the warmer and cooler chambers. In the IVDS, the supersaturation achieved is such that particles smaller than 10 nm can be activated and counted.

2.3   IVDS-BETA

2.3.1   IMPROVEMENTS AND TESTING

Many improvements were made to the IVDS-Beta instrument over the course of several years and they are discussed in this section. The IVDS-Beta had been in use for more than 6 years and remained available for use for many years after being "retired" because it had proven to be a reliable and dependable device for the detection of viruses. Initial improvements focused on allowing lower concentrations of viruses to be detected and counted. These were successful and are described below. Other improvements were made to make the instrument more robust, easier to use, and more sensitive and operator friendly. All of these improvements were incorporated into the second version designated the IVDS-Com.

Capitalizing on the physical characteristics, we found it possible to separate virus families and count the individual viruses in a new and dramatic way using easily obtained materials and simple techniques. The only materials used in the IVDS are a buffer, CO_2 gas, and butanol (the butanol-based CPC was later replaced with a unit that required only water). These early upgrades represented advances in the concentration and counting phases. Results indicated a practical, easy to use device capable of counting viruses in near-real time. The final configuration is expected to analyze samples, count the viruses present, and yield preliminary identifications for all viruses.

The Defense Advanced Research Project Agency (DARPA)-funded work on the IVDS-Beta yielded substantial advances in development of the Gas Phase Electrophoretic Mobility Molecular Analyzer (GEMMA) by incorporating a number of modifications and upgrades to its detector unit. Laboratory studies further validated the IVDS as a feasible solution for providing near-real time, stand-alone monitoring for all viral threats, known or unknown.

Work focused on the machining and integration of a second-generation, three-module ultrafiltration (UF) stage and the testing the tandem UF-GEMMA system's ability to purify and detect viruses at various concentrations and levels of background impurities. A density gradient ultracentrifugation stage was investigated and plans were made to integrate it into the IVDS. The UF and GEMMA stages were expected to be greatly simplified after the centrifuge was integrated. A range of impurity levels were investigated in ongoing studies (up to 0.5% cesium chloride and 1.0% thyroglobulin) to ensure that the worst possible cases were addressed. The results indicated that the

- Second and third modules of the UF performed well for retention volumes down to 10 µL.
- First module showed excellent purification performance even for large proteins but needed modification to minimize loss of viruses.
- Improvements made on the GEMMA, which shows superb performance when vapor-lock is avoided, appear to be bringing the system to operational consistency.
- Current sensitivity of 107 virus particles (between 103 and 107 plaque-forming units [pfu]) may be improved to fewer than 103 pfu (sensitivity of PCR) by the substitution of a new CPC and a new ES unit into the GEMMA.

2.3.2 BACKGROUND

The IVDS-Beta project began with the development of the theory and concepts needed to build a device capable of analyzing virus particles which are among the smallest known microorganisms. Biochemical techniques were excluded because they are inherently limited in the number and types of viruses and virus-like particles they can analyze in a short time. The physical characteristic methodology was found to be generic in nature, quick, and unlimited in its ability to count and analyze all viruses. Experiments to date have advanced this early work and now IVDS can routinely examine such viruses as MS2. Further advances have been made in the concentration and counting stages.

2.3.3 SYSTEM DESCRIPTION

The IVDS-Beta simplified detection and monitoring of submicron particles or macromolecules, particularly viruses and virus-like materials in biological fluids, water, air, and other fluids. The unlimited detection of viruses of any family, genus and species was an extremely challenging problem for current technology. IVDS was a breakthrough development and provided accurate identification and analysis (including sample concentration and purification) in less than 15 minutes for most samples.

A virus sample is purified and concentrated via a UF process. The UF module allows the removal of interfering proteins and other biological entities while retaining the virus species for further study. The counting stage (GEMMA unit) uses an ES nozzle, DMA, and a CPC.

2.3.3.1 Ultrafiltration (UF) Module

The first module of the three-module UF stage is a hollow fiber-based tangential flow filtration system. The system operates by pumping the feed stream through a hollow fiber (Figure 2.6). As the solution passes through the fiber, the sweeping action of the flow helps prevent clogging. A pressure differential forces the filtrate through the fiber while the virus feed stream is purified and concentrated. A wide range of filter pore sizes is available. This filtration technique can reduce volumes from 5 to 0.2 mL. The first module was assembled and tested successfully, as reported in more detail in Chapter 5.

The second module is shown in Figure 2.9. It is a flat-membrane diafiltration apparatus that reduces volume from 0.2 mL to 400 to 10,000 nl. To our knowledge, no other UF device with a retention volume of this order of magnitude has been built.

The third module that accepts output from the second module and inputs it to the detector was also developed and tested successfully. This module can also accept outputs from other sample feed streams. For example, capillary sample holders would allow nanoliter-size samples to be transported by mail to a centrally operated GEMMA unit.

A key feature built into the overall UF design is the independence of the three modules and it has both performance and practical implications. One performance implication is that the first module can, after passing its output to the second module, move along to processing the next sample while the second module is still processing the first. Likewise, the second and third modules can operate this way, thus reducing total operating time. Maintenance is also easier because replacement of a single module allows the total system to be brought back on line rapidly.

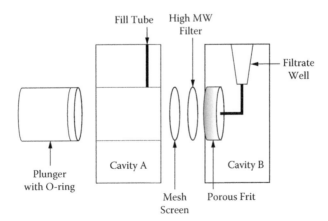

FIGURE 2.9   Flat membrane diafiltration apparatus.

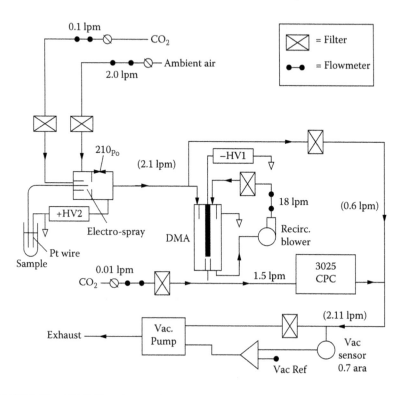

FIGURE 2.10   GEMMA unit operation.

2.3.3.2   GEMMA Unit

The GEMMA Unit is composed of a sample input assembly, electrospray module, differential mobility analyzer, condensate particles counter, and controlling computer and software. The integral parts are described below. Figure 2.10 is a flow schematic.

2.3.3.2.1   Sample Input Assembly

The inlet tube for the GEMMA is a fused silica capillary of 25 µm inside diameter and 150 µm outside diameter. The capillary is inserted into a section of polyetheretherketone (PEEK) tubing of appropriate inner diameter and the PEEK tubing is swaged into the front orifice of the electrospray unit. The sample is aspirated into the electrospray module by vacuum generated by a pump that meets appropriate flow parameters for sample passage through the GEMMA.

2.3.3.2.2   Electrospray Module

The electrospray module subjects a conductive liquid to a strong electric field. The field produces a cone that emits a fine jet that then breaks up into small droplets and forms a fine plume. In the electrospray unit of the GEMMA, the droplet electrical charge is reduced to a single charge or is neutralized by an alpha particle emitter that ionizes the surrounding air. To eliminate the possibility of the breakdown (corona discharge) of the air in the plume caused by the high electric field, the spray tip is surrounded by a flow of CO_2.

An improved electrospray apparatus was obtained from TSI, Inc. of St. Paul, Minnesota, and integrated into the front end of the detector. The new electrospray positions the inlet capillary closer to a smaller aperture plate. These spatial improvements allow a greater percentage of the sample stream to be processed by the detection hardware. Several other internal improvements of the electrospray unit were also made and increased system sensitivity tenfold.

2.3.3.2.3 Differential Mobility Analyzer (DMA)

The DMA separates particles by their electrical mobility in air. The sample stream flows through a gap between a rod and a cylinder enclosing an electrical potential. Particle mobility, which is related to size and charge, will either pass particles through the DMA or cause them to impinge on the walls. Mobility for singly charged particles generated by the electrospray becomes a direct measure of particle size.

2.3.3.2.4 Condensation Particles Counter (CPC)

In the CPC, the sample particles flow in tandem with a saturated butanol working fluid. The nano-sized particles initiate the condensation of the butanol and the stream is then cooled. A standard optical counter then counts the butanol-condensed particles and the results are displayed by the computer software. Replacement of the original Model 3025 CPC component with a Model 3010 increased sensitivity thirtyfold.

2.3.3.2.5 Computer Software and Hardware

The GEMMA was controlled with a 486 personal computer utilizing the Scanning Mobility Particle Sizer (S1/1PS) software developed by TSI. A number of software options are available for displaying and interpreting the data. Particles are classified by one of the following values:

- Number
- Surface area (assuming spherical shapes)
- Volume of particles (assuming spherical shapes)

The units displayed by the S1/1PS can be set up to display the following measurements:

- Counts—raw data without corrections
- Cone dW/dlgDp—normalized concentration (useful for comparing data from other instruments)
- Cone dW—concentration within particle size bin
- Log dW/dlgDp—vertical scale compressed to log10
- Cone (W%)—concentration as percentage of total concentration
- Cumulative W—particle concentration in cumulative format
- Cumulative W%—particle concentration in cumulative percentage format

Several options are available for data display, for example, graphs or tables. Data may also be exported into a spreadsheet program.

2.3.4 UPGRADES

A number of improvements to the IVDS-Beta enhanced its performance. The electrospray upgrade to the GEMMA unit increased the sensitivity of the detector approximately tenfold. A pressurized chamber to help eliminate capillary plugging was purchased. A new CPC improved the detection limits of the GEMMA unit thirtyfold.

A pressurizable chamber (Figure 2.11) purchased from TSI holds the GEMMA sample input container (third UF module). This chamber allows positive air pressure (a few pounds per square inch) to be applied to the sample liquid. This was a simple and effective way to eliminate bubbles that could have vapor locked the GEMMA input capillary. A shortening of the detector capillary input was implemented to greatly reduce the lag time between measurements and reduce complications from extremely large viruses.

Pump and auto sampler methods were identified for incorporation. In addition, high molecular weight filters became available for the second UF module and

FIGURE 2.11   Sample holding chamber that uses pressure to push sample into GEMMA unit.

minimize absorptive losses of viruses in the filter. Further improvements to this device are discussed in Chapter 4.

The MS2 bacteriophage was found to be an excellent virus standard. It is stable in solution and is not affected by dilutions. MS2 is well characterized in the literature and its 26-nm size works well with GEMMA analysis procedures. The other virus tested, an *Escherichia coli* T7 bacteriophage, was analyzed and made available as another size standard.

The GEMMA unit was extremely reliable in counting small quantities of viruses and continues to present a high signal-to-noise ratio around virus peaks. In addition, the GEMMA analysis of particle size standards obtained from Duke Scientific generated results that were very close to the claimed size distributions. The UF modules were tested by the successful removal of 1% thyroglobulin from a solution while incorporating the appropriate buffer for GEMMA analysis.

2.3.5 TESTING OF IMPROVEMENTS

The laboratory work focused on testing the UF-GEMMA tandem system's ability to purify and detect viruses at various concentrations and levels of background impurities. Several standard virus samples of known pedigrees were analyzed and used as markers in the purification and concentration studies that followed. A range of impurity levels was investigated (up to 0.5% cesium chloride and 1.0% thyroglobulin) to ensure that the worst possible cases were addressed. Also tested were polystyrene sphere standards of known particle sizes to determine the accuracy of the sizing of unknown samples.

2.3.5.1 General Test Protocols

To process a sample through the GEMMA analyzer, the following steps should be considered as a guide. Because most samples processed contained varying degrees of contamination, orders of magnitude differences in virus concentration were artificially produced as samples for laboratory studies. Figure 2.12 is a general decision tree for processing samples in the IVDS. Every sample received had to be judged "clean enough" to be analyzed by the GEMMA. The criteria for GEMMA analysis are as follows:

- Interferences from biological contamination must be low enough to allow GEMMA analysis. High protein concentrations will swamp the detector display and cover any virus signals of interest. If a sample's protein concentration is deemed too high, it must be processed through module 1 of the UF apparatus.
- Any salt solution contamination at high concentration will produce the same masking of the virus signal of interest produced by high protein concentrations. A high salt concentration also requires processing through module 1 of the UF apparatus.

In addition, the amount of virus contained in s sample must be high enough to be counted when the sample is aspirated into the GEMMA detector. The easiest way to

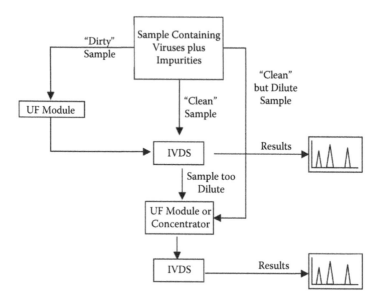

FIGURE 2.12   General decision tree for processing samples in IVDS.

determine the sufficiency of a virus for counting is to run a GEMMA analysis on the sample. The analysis is easy to perform and can be completed in a few minutes. The following protocols were used as general guides for the analysis of virus-containing samples in studies.

2.3.5.1.1   Protocol I

- Determine whether the sample is clean enough for direct processing in GEMMA.
- A clean sample can now be analyzed in the GEMMA detector and results can be printed or archived.
- A sample that is too dilute will require further concentration in module 2 of the UF apparatus. Module 2 can reduce the volume significantly to allow successful analysis and interpretation of the virus sample.

2.3.5.1.2   Protocol II

- A sample that is too dirty for direct processing in GEMMA will require preanalysis processing.
- A contaminated sample should be processed in module 1 of the UF apparatus by adding enough buffer in the filter loop to replace any interfering species. Typically, samples were processed in module 1 with a 40:1 (buffer:solution) volume addition. The 40:1 addition removes the contaminating species and allows sufficient concentration of the virus sample to enable processing by the GEMMA detector.

- The clean sample can now be analyzed in the GEMMA detector and results can be printed or archived.
- A sample that is too dilute will require the further concentration in module 2 of the UF apparatus. Module 2 can reduce the volume significantly to allow successful analysis and interpretation of the virus sample.

2.3.5.1.3 Protocol III

- If a sample is known to be clean but not sufficiently concentrated, it will require preanalysis processing.
- The clean dilute sample can be placed directly into module 2 of the UF apparatus and concentrated in one pass through the module. Typically the sample will be concentrated to the fullest extent possible by the module. However, if the sample should become too concentrated and possibly plug the inlet capillary of the GEMMA, it can be diluted back with a buffer solution.
- The clean sample can now be analyzed in the GEMMA detector and results can be printed or archived.

2.3.6 Processing and Analysis of MS2 from Cell Culture

One key step in this work was to obtain a highly purified preparation of MS2 that could be used to insert known amounts of virus into samples containing standardized background components. MS2 preparations obtained from all sources contained high levels of background components, including proteins (including very high molecular weight proteins), bacterial pili and fragments thereof, and salts, all of which create interfering signals in the high-sensitivity GEMMA detector.

In the process of producing highly purified MS2, some very interesting results were seen from the partially completed IVDS that indicated the system's capabilities surpassed prior estimates. Some of the tests involved the analysis of MS2 bacteriophage following a purification and detection protocol that was more similar to the protocol developed for the final IVDS than any sample analyzed to this point. Although the IVDS density gradient centrifugation stage was not yet online, the MS2 sample obtained for these tests was purified by density gradient centrifugation before we received it. An additional 300-fold increase in sensitivity was achieved by the integration of a new electrospray apparatus and a different model CPC.

Purified samples were dialyzed repeatedly in an ammonium acetate buffer because this buffer is virtually invisible to the GEMMA detector. Dialysis removed all remaining traces of non-volatile salts (phosphate, cesium chloride, etc.) and other soluble impurities such as proteins likely to produce "salt peaks" at 10 nm or less. Any peaks above that level are created by actual viruses, aggregates, viral fragments, and non-viral impurities ranging in size from 10 to 800 nm.

2.3.6.1 Standards and Specifications

We used certified nanosphere size standards purchased from Duke Scientific Corp. as reference particle size standards in this study. The standards are monodispersed polystyrene microspheres calibrated with methodology developed by the National Institute of Standards and Technology (NIST). The standard diameters were 73 and 54 nm.

2.3.6.2 Cultures Tested

The cultures tested from known sources were an antigen grade MS2 bacteriophage and an *Escherichia coli* T7 bacteriophage purchased from the American Tissue Type Culture Collection (ATCC). The antigen grade MS2 was at a concentration of 1×10^{13} pfu/mL on arrival and was density gradient centrifuged and dialyzed. The T7 bacteriophage arrived in a freeze-dried state at a concentration of 2×10^8 pfu/mL. The freeze drying also retained the growth media composed mainly of skim milk products.

2.3.6.2.1 MS2 Bacteriophage

Previously analyzed MS2 samples showed evidence of high dimerization tendencies. It is well known that electron microscopy reveals MS2 particle diameters of 24-27 nm. However, what is perhaps less known is the arrangement of MS2 virions in pairs on bacterial pili, like barbells, as shown in TEM micrographs.

The micrographs suggest why we might see a size corresponding to an MS2 dimer in the GEMMA detector data. If the viruses remain complexed to the pili when the structure reaches the electrospray, the electrospray action may break a pilus along its length, freeing viruses in pairs. In previous work, very long molecules, particularly those containing weak linkages, have been shown to break up along their length in the GEMMA electrospray, yielding sizes corresponding to the fragments.

In previous work on MS2 (Wick et al. 1999a), what we interpreted as a dimer peak always appeared between 37 and 41 nm. The dimer does not appear at a particle size of 48 to 52 nm (size for two attached monomers) because the particle size algorithm in the CPC fits measured sizes into an apparent spherical diameter. This fitting algorithm yields the 37 to 41 nm depicted in the graphs. At least one virus exhibited evidence of four peaks corresponding to monomer, dimer, trimer, and tetramer, with the peak positions (diameter D) fitting well to a formula where the exponent is 0.39:

$$D = DO \, N^a$$

We should expect the exponent to fall between 1/2 and 1/3, corresponding to cylindrical-like and spherical-like growth with aggregation number N, as observed.

2.3.6.2.2 T7 Bacteriophage

The T7 bacteriophage is described as a P-22-like phage. The head is approximately 60 nm in diameter and icosahedral in symmetry. The tail to which six short fibers are attached is approximately 17 nm long and 6 nm wide.

2.3.7 TEST MATRIX

Table 2.1 presents an overview of the tests performed using the ultrafiltration modules and the GEMMA analyzer.

TABLE 2.1

Overview of Tests Using Ultrafiltration Modules and GEMMA

kV	nA	Sum of ROI Counts (22.5–27.9 nm)		
1.75	−228	1457.4		
1.80	−232	1458.5	Count average	
1.85	−235	1454.9		1438
1.90	−238	1408.9	Standard deviation	
1.95	−243	1446.1		28
2.00	−249	1430.9		
2.05	−254	1437.6		
2.10	−260	1411.9		
2.15	−264	1392.1		
2.20	−269	1404.9		
2.25	−274	1465.4		
2.30	−280	1483.3		

2.3.8 TEST RESULTS

2.3.8.1 MS2 Bacteriophage Virus Standard

The MS2 bacteriophage was chosen as a standard for its stability and uniform size. Although the virus was received as a dimer because the culture was fresh, the dimers quickly separated into monomers and stabilized in solution. This section shows the results of tests with MS2 virus concentrations between 1×10^{13} and 1×10^8 pfu/mL.

2.3.8.1.1 Undiluted and Tenfold Diluted MS2 Samples

Figure 2.13 is a scan of the full-strength sample that after the dialysis protocol was at a very high concentration of 1×10^{13} pfu/mL (that attests to the effectiveness of the purification and concentration protocol); the salt peak is centered at 10 nm. A small peak is visible at 26 nm, with a dominant peak at 37 nm. A closer view of the MS2 signal is presented in Figure 2.14, which spans the 19–45 nm region.

The sample was then diluted tenfold and analyzed again. The effects of dilution are different for salt peaks and "real" particles. A tenfold dilution shifts the position of the salt peak by a reduction of $10^{1/3} = 2.15$, in this case from 10 to about 5 nm, thus increasing the signal-to-noise ratio in the region of 10 to 20 nm. For a real particle such as a virus or viral fragment, the position remains approximately fixed but the intensity is reduced approximately tenfold. The resultant scan is presented in Figure 2.15.

2.3.8.1.2 Analysis of Results

Because of the increase in signal-to-noise ratio in the 10 to 20 nm range, Figure 2.15 reveals some small features that are not present in the undiluted samples shown in Figures 2.13 and 2.14. There is some noise in this range, but previous experience with MS2 and RNA allowed us to look in certain areas for information on viral

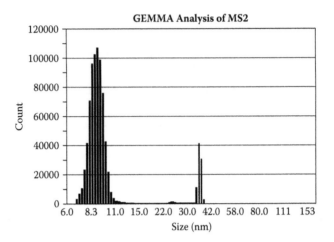

FIGURE 2.13 MS2 sample at full strength. After dialysis protocol, concentration was 1×10^{13} pfu/mL.

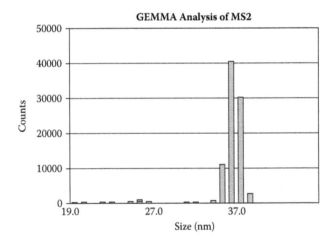

FIGURE 2.14 Closer view of MS2 signal in Figure 2.13 that spans 19 to 45 nm region.

fragments. In particular, as exemplified in Figure 2.16 for a 2-year-old preparation of MS2, we generally see a peak at 13 nm that we believe corresponds to a protein-aceous subunit of MS2.

The MS2 coat protein (PDB 1MSC protein [Protein Data Bank]) which comprises 69% of the viral mass is known to dimerize (form dimers) that x-ray analyses indicate have lengths of 12.9 nm. These dimers then assemble into units of five known as pentamers, also of 12.9 nm length. The viral coat is composed of 12 such pentamers plus an additional 12 dimers, to form an icosahedral structure. Clearly, based on a 26 nm diameter (i.e., 13 nm effective radius) of the entire particle, and the coat consisting of 69% of the mass, all three dimensions of the pentamer must be approximately 13 nm and should thus be registered in the GEMMA at ~12 or 13 nm.

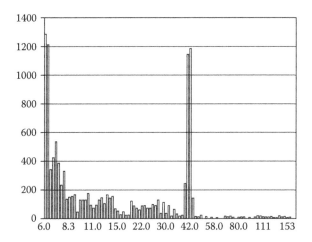

FIGURE 2.15   Tenfold dilution of sample in Figure 2.14 that shifts salt component.

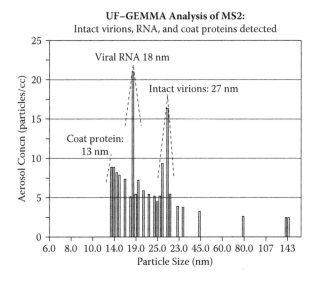

FIGURE 2.16   Analysis of 2-year-old preparation of MS2.

Because the virus forms by assembly of the pentamers, it is reasonable to assume that it may well disassemble into the same units to a large extent. In view of the fact that aged virus preparations have shown increased peak intensities at 12 to 13 nm with qualitative correlation of age and intensity, we believe that the 12- to 13-nm intensity is probably a sign of fragmentation. Its absence in fresh MS2 preparations indicates that it is not created by breakup in the electrospray. This is supported by experiments on several protein complexes that also remained intact.

The GEMMA size measured for the RNA of MS2 can be estimated accurately from published reports of ssDNA analysis using the device. In the formula derived from the work of Mouradian et al. (1997) for ssDNA and RNA, X represents the number of nucleotides:

$$D = 1.11\ X^{1/3}$$

The MS2 genome is 3,569 nucleotides in length and thus D = 17 nm. The formula was derived for rather short ssDNA strands, and we found in previous work that aged MS2 shows a peak closer to 18 nm that we attribute to viral RNA.

Within the limitations set by the noise in the above data, we can deduce evidence from Figure 2.16 for a small protein subunit peak centered at 13 nm and an equally small RNA plot centered at 18 nm along with the monomer and dimer virus peaks at 26 and 37 nm. The fragments, protein subunit, and RNA, are clearly an order of magnitude lower in intensity (and thus concentration) than the monomer and dimer intensities, indicating a low degree of fragmentation under the conditions applied. It should be noted that even in the diluted sample, any result below 10 nm is meaningless because of the entrance of soluble proteins at and below that level.

2.3.8.1.3   Analysis of Signal-to-Noise Ratio

For a quantitative analysis of the signal-to-noise ratio shown in Figure 2.14, we must examine the raw data because it provides a far more accurate picture than visual inspection of the graphs. Table 2.2 shows the detector data.

As shown in Table 2.2, there are about 40,000 counts at the main peak channel, 36.5 nm (corresponding to virus dimers), and a total of 80,000 counts over the three-channel spread of the peak. What was not accurately shown in the graph but is evident from the tabulated data is the extremely low background just outside the peak region. Indeed, in the larger particle size direction, only three channels beyond the peak, the intensity drops to literally zero counts per channel, and four of five consecutive channels show zero counts per channel. This indicates two results:

- The resolution of the peak is very sharp.
- The signal-to-noise ratio is extremely high, far higher than the 2,000:1 previously estimated.

2.3.8.1.4   Results of Further Dilutions

A concentration of 2×10^{10} pfu/mL was sampled straight into GEMMA, yielding fairly high counts (several hundred per channel at the peak position), and a signal-to-noise ratio of about 25:1, as shown in Figure 2.17. Note that the virus concentration and the GEMMA amplitude are about 2 to 2.5 orders of magnitude lower than those from the 1×10^{13} pfu/mL sample analyzed earlier that showed about 40,000 counts per channel at the peak.

After diluting an additional 100-fold, a 2×10^{8} pfu/mL solution was sampled directly into the GEMMA. This low concentration of virus was detected with a small number of counts, but still maintained an exceedingly good signal-to-noise ratio, as shown in Figure 2.18. Note how the salt peak due to remaining soluble impurities has been pushed off the chart axis, which extends below 3 nm.

TABLE 2.2
Raw Data for Signal/Noise
Ratio Analysis of MS2

Channel Particle Size Midpoint	Gemma Counts
23.7	87
24.6	364
25.4	877
26.4	269
27.4	49
28.4	66
29.4	54
30.5	116
31.6	107
32.8	75
34.0	533
35.2	10,800
36.5	40,359
37.8	30,052
39.2	2571
40.7	77
42.1	0
43.7	33
45.3	0
47.0	0
48.7	0
50.5	15
52.3	0
54.2	0
56.2	14
58.3	0
60.4	0

Figure 2.18 demonstrates the feasibility of detecting viruses at low concentrations and very small number counts, with levels of purification that yield zero counts in most adjacent channels. Thus, the 1990s work that started with MS2 viruses buried in a bacterial culture and a combination of density gradient centrifugation, ultrafiltration, and a GEMMA device detected the MS2 virus with a 15,000:1 signal-to-noise ratio. Such high ratios are rare in experimental science for any experimental probe, particularly when the starting point is a virus lying within a sea of biological background material.

An interesting addition to these findings is an example of counting a sample of MS2 concentrated at 10^8 pfu/mL. It was an easy matter to count the samples over a longer time and accumulate counts over hundreds of minutes to improve detection

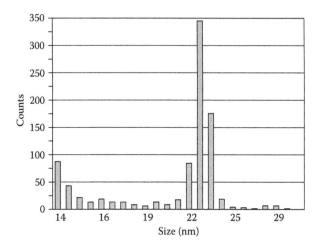

FIGURE 2.17   Signal-to-noise ratio of about 25:1.

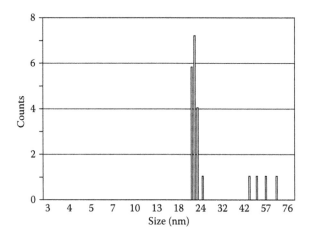

FIGURE 2.18   Additional 100-fold dilution of sample in Figure 2.17.

limits when utilizing a particle counting technology. For example, seawater was analyzed and counted for 100 minutes, representing a 10^2 improvement in detection
limits. This means a 10^8 pfu/mL sample could be diluted to 10^6pfu/mL and produce
the same results.

It is also a simple matter to concentrate a sample further, usually 100 to 1,000
times concentration range to improve the level of concentration that can be achieved
further to the range of 10^4 to 10^3 pfu or more should the need arise. Improvements
in the sampling of the CPC and other hardware advancements can be expected to
further improve concentration limits by another 100 times or more. In practice, such
low dilutions are unnecessary because the virus loading is usually high enough to
run routine samples within the 10^4 to 10^6 pfu/mL range.

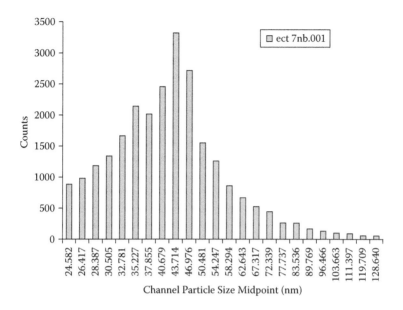

FIGURE 2.19 IVDS processing of T7 bacteriophage showing 3000 counts per channel.

2.3.8.2 T7 Bacteriophage

In addition to the MS2 phage used in this study, we also obtained a sample of *Escherichia coli* T7 phage from ATCC. This was reconstituted as a 2×10^8 pfu/mL solution, according to the instructions provided with the material. The solution was then diluted tenfold to 2×10^7 pfu/mL and run into the GEMMA.

The T7 phage was clearly detected at 3,000 counts per channel, as shown in Figure 2.19. It is interesting to note that according to the formula derived above for converting counts to particle concentrations, we estimate that the concentration of virus particles was about 10^{10} particles/mL, whereas according to the information supplied by ATCC, the concentration should have been 2×10^7 pfu/mL. This means that a single plaque-forming unit corresponds to about 1,000 virus particles. Whether this figure was inflated due to overly conservative estimates of the virus potency by ATCC should be determined in the future as the interesting issue of pfu–virion conversion comes to the foreground.

2.3.8.3 Particle Size Standards

Nanoparticle standards of average diameter 73 nm (±2.6 nm) and 54 nm (±2.7 nm) were obtained from Duke Scientific Corporation and analyzed by the GEMMA. At a concentration around 10^{12} particles/mL corresponding to 0.05% by volume, the nanoparticles were detected easily and sized in excellent agreement with the quoted size distribution. Approximately 1,400 and 4,000 counts were obtained in the peak region for the 73 and 54 nm standards, respectively, as shown in Figure 2.20 and Figure 2.21.

This analysis of standardized particles allowed us to develop a formula for converting peak counts to particle number density: the concentration in particles per

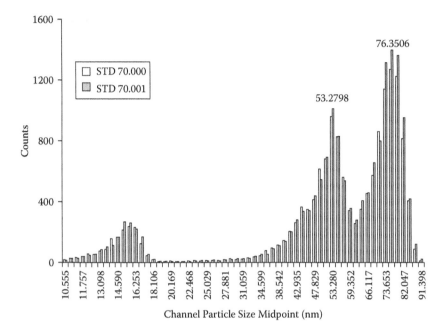

FIGURE 2.20   Approximately 1400 counts are seen for 77 nm standards.

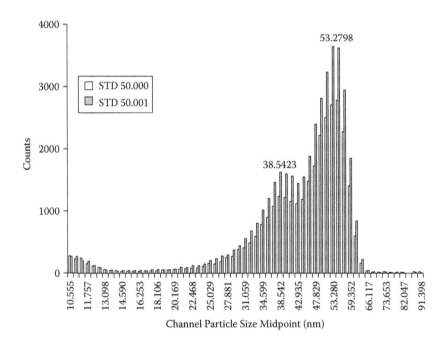

FIGURE 2.21   Approximately 4000 counts are seen for 54.0 nm standards.

milliliter = 2×10^8 (total number of counts in peak) under current operating conditions normalized to a 60 second run time. This sets a GEMMA lower detection limit under current conditions of approximately 10^9 particles/mL in a UF-concentrated output. If the output volume of the second module of the UF is 10 μL, the lower limit of detectable particle number count (as opposed to number concentration) is about 10^7 particles.

In the worst case, 1 pfu equals 1 virus particle, but in general 1 pfu will contain many virus particles. Indeed, for the case of the T7 virus, it is likely that 1 pfu equals several thousand virus particles. Therefore the detection limit of 10^7 virus particles falls between 10^3 pfu and 10^7 pfu, with 10^5 pfu probably typical. With the new CPC system discussed earlier, these detection limits could be reduced readily below 10^3 pfu, thus rivaling or exceeding results from the highly capable polymerase chain reaction (PCR) methods. Note that PCR methods handle only specific viruses for which sequencing data and primer molecules are available.

2.3.8.4 Standardized Salt Solutions

Cesium chloride solutions, with concentrations starting at 0.5% and proceeding downward in decades, were analyzed directly by the GEMMA without purification. The resultant scans are illustrated in Figure 2.22 and Figure 2.23. The position of the resulting salt peak showed a cube root dependence on concentration, up to 0.05%, the relocation of the salt peak with a change in concentration. For the sample at 0.5%, the effect of the salt concentration on conductivity of the solution and thus operation

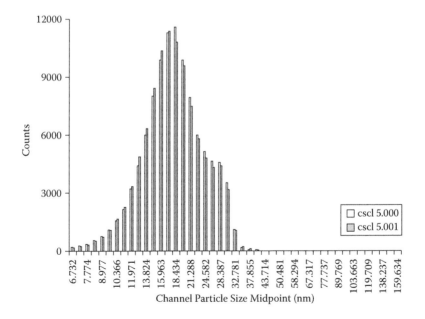

FIGURE 2.22 Second filtration module yields improved signal-to-noise ratio, with amplitude dropping from ~1000 counts per channel to 0 only five channels away from peak center.

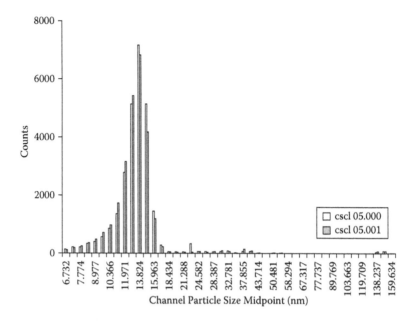

FIGURE 2.23 Further reduction to about half the retention volume of that shown in Figure 2.22.

of the electrospray was seen easily as a departure from the cube root position and a perturbation in the shape of the peak.

Assuming that the electrospray operates normally at 0.05% CsCl, we can calculate the droplet size from the 13 nm peak position. Taking into account the 3.99 density of solid (dried, in-flight) CsCl, the droplet size is calculated at 265 nm, which fits very well with the expected size on the order of 300 nm according to the operating conditions. The 265 nm figure will be used in subsequent computations for converting salt peak positions and impurity concentrations.

2.3.9 ULTRAFILTRATION AND GEMMA TANDEM ANALYSIS OF MS2

Beginning with the highest starting concentrations of virus and working down to the lowest, we present results of the analysis of density gradient, centrifuge-purified MS2 with the newly constructed UF device, module 2. The samples of MS2 were highly purified and thus could be processed directly into module 2. Figure 2.24 and Figure 2.25 show, respectively, the GEMMA response before and after application of module 2 utilizing a 1MDa filter; the starting concentration of MS2 was 2×10^{13} pfu/mL.

After one compression in the second module, the ratio of peak amplitude to the amplitude at the highest noise channel (at 9.5 nm) increased from ~2:1 to ~6:1 with ~18,000 counts in the highest peak channel. Note that without diafiltration, the number of counts in the salt peak (centered at 9 nm, approximately 2,000 counts in the highest channel) did not change.

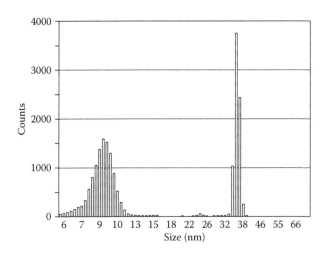

FIGURE 2.24 GEMMA results before ultrafiltration.

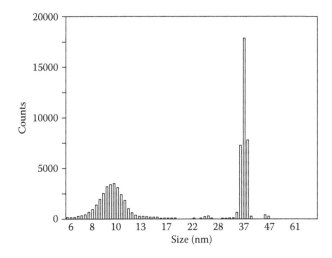

FIGURE 2.25 GEMMA results after ultrafiltration.

This underlines the importance of successive diafiltrations. Simply pushing solution through the UF membrane will reduce the quantity of retentate liquid and numbers of non-viral impurities greatly, the actual concentrations of these impurities will not change. The impurities can be reduced only by diafiltration in which clean buffer is added to dilute the impurities before filtration steps. In any case, virus concentration is increased by a simple filtration without diafiltration examined at the 37 nm position, as seen in Figures 2.20 and 2.21.

The increase in the peak amplitude at 37 nm after the compression is about fivefold, from 3,600 to 18,000 counts per channel. This correlates well with the approximate fivefold reduction in volume achieved in the second module step, namely reduction of the volume from ~200 to 30 to 40 μL. Whenever a fivefold reduction in volume yields a fivefold increase in signal (in turn indicating a fivefold increase in concentration), the virus retention by the UF module is very high; nearly all the virus is retained by the membrane and is present in the extracted liquid.

Starting with an MS2 concentration of 2×10^{11} pfu/mL and applying one compression in the second module yielded a superb signal-to-noise ratio, with amplitude dropping from approximately 1,000 counts per channel to literally zero only five channels away from the peak center. This occurred on both sides of the peak, as shown in Figures 2.22 and 2.23. The difference between the two plots is that the second module compression carried out in the Figure 2.23 plot was about half the retention volume shown in the Figure 2.22 plot, yielding ~20 instead of 40 μL. The doubling of the signal between the two conditions indicates that the GEMMA peak amplitude is roughly linear with final retention volume. This linearity again supports the high degree of retention of viruses.

Note that the peak at 23 nm (MS2 monomer) assumed dominance over the dimer peak at 37 nm on the initial sampling of the 10^{13} pfu/mL MS2 bacteriophage, Later sampling of the virus with lower concentrations over several months indicated the dimers separated into monomeric virions. Processing 200 μL of a 2×10^{10} pfu/mL MS2 solution through module 1 resulted in a sample of only 4×10^9 pfu. This was injected into the second module and after four diafiltrations, an excellent signal-to-noise ratio was obtained. The results from this sample are presented in Figure 2.26.

With a peak amplitude of over 100 counts per channel, the amplitude drops to a few counts per channel only three channels to either side of the maximum. This demonstrates a signal-to-noise ratio of about 25:1 and a size resolution of ±1 nm. The facts that four diafiltrations were performed and 100 counts per channel were obtained at this input concentration means that the second module was able to retain a high fraction of the viruses throughout five compression steps.

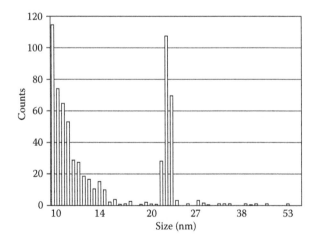

FIGURE 2.26 Sample processed through second module. After application of four diafiltrations, an excellent signal-to-noise ratio was obtained.

Processing 250 μL of an 8×10^7 pfu/mL MS2 solution through module 1 yielded a starting amount of only 2×10^7 pfu. This sample was fed into the second module and produced a very good signal-to-noise ratio after one compression. The amplitude dropped from 200 counts per channel in the peak area to 7 or fewer five channels to either side of the highest peak channel, making the signal-to-noise ratio better than 25:1.

2.3.10   Ultrafiltration and GEMMA Tandem Analysis of MS2 with Thyroglobulin

A 1 wt% solution of the large thyroglobulin protein with 5×10^{10} pfu of MS2 added was processed through both modules of the UF. Module 1 was equipped with a 750 KDa cross flow filter cartridge and Module 2 was fitted with a 1 MDa filter disk. The resulting GEMMA analysis in Figure 2.27 shows a virus peak although the intensities were somewhat lower than expected based on previous experiments. One would expect more on the order of 100 counts per channel in the peak. Due to the low count rate, it appears likely that the first module caused some loss or fragmentation of the virus under the conditions applied.

The removal of thyroglobulin by the first module was excellent. A 1% concentration of any soluble (and non-volatile) impurity should give rise to a broad salt peak centered around 60 nm. The salt peak in this experiment lies at 10 nm (with a low intensity of fewer than 400 counts). A tenfold reduction in the center position of the salt peak indicates (because of the third-power dependence of the volume of a sphere on the diameter) a thousandfold reduction in the concentration of the impurity.

Thus the UF reduced the thyroglobulin concentration from 1% down to 0.001% or less. It should be mentioned that no known malfunction of the GEMMA could

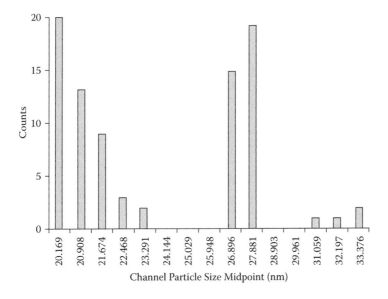

FIGURE 2.27   Thyroglobulin, with 5×10^{10} pfu of MS2 processed through IVDS.

conceivably give rise to a salt peak at only 10 nm when the concentration of any soluble impurity significantly exceeds 0.001%. The thousandfold reduction of thyroglobulin concentration is quite remarkable, considering the following:

- The molecular weight (MW) of the protein is only slightly smaller than the rated MW of the hollow fiber used in the first module (750 KDa). The company that supplies the 750 KDa low internal volume hollow fiber cartridge later produced a 1 million D cartridge that will yield much better performance with large proteins.
- The second module of the UF probably did not add a lot of activity to the removal of the protein; most of the thousandfold reduction came from the first module.
- The appearance of the virus peak as shown in Figure 2.27 demonstrates that at least a portion of the virus is being retained even as the large protein is almost completely removed. Studies are ongoing to extend this type of result to a wider range of proteins.

2.3.11   SUMMARY OF CLASSIFICATION TESTING

- The MS2 bacteriophage, grown in a bacterial culture and rich in biological impurities was purified by density gradient centrifugation, followed by dialysis (very closely related to ultrafiltration) in a purification protocol simulating that envisioned for the final IVDS design. This method purified and effected a tremendous concentration of the virus.
- After purification and concentration with the gradient centrifugation and dialysis protocol, the MS2 bacteriophage yielded a huge GEMMA detector signal of 40,000 counts per channel and a signal-to-noise ratio better than 10,000:1.
- In addition to intact virus particles, we found evidence of viral fragments and viral RNA in the detector data, yielding at least semiquantitative data on the degree of virus fragmentation. Furthermore, the degree of complexation of the virus was dramatic in that the peak corresponding to MS2 dimers was an order of magnitude larger than that for single virus particles for freshly prepared MS2 samples. This dimerization broke down over a month in storage to yield a high fraction of monomeric MS2. The dimerization can be better visualized by reference to electron micrographs in the literature that show MS2 prepared from bacterial cultures arranged in pairs on the bacterial pili, like barbells (Wick et al. 1999).
- A solution of 2×108 pfu/mL was detected with a small number of counts and a dramatic signal-to-noise ratio. Most channels outside the virus peak region registered zero counts.
- Ultrafiltration successfully removed a large thyroglobulin protein at a concentration of 1% in a 2×10^9 pfu/mL MS2 solution and yielded a good signal-to-noise ratio despite some loss of virus due to the action of the first module.

- Standardized particles of 74 and 54 nm diameters and known concentrations were obtained and used to calibrate the GEMMA for size and concentration measurements.
- Using standardized salt solutions, the linearity of the detector was determined (along with electrospray droplet size) along with the point at which linearity broke down due to a change in liquid conductivity.
- An *E. coli* T7 bacteriophage was established as a second standard virus for use in IVDS evaluations. Interestingly, our work indicates that a single plaque-forming unit (pfu) obtained from ATCC may contain as many as 1,000 virus particles.

2.4 COMMERCIAL IVDS

The next generation of IVDS incorporated many improvements to the original model. The major changes included separation of the components and simplifying upgrades. The CPC was taken out of the cabinet and likewise the DMA was positioned outside the electronics housing. This allowed the use of multiple DMAs with different sensitivities. The electrospray assembly was placed alone into a box. The connection between the mechanical cabinet (air flow, filters, and the like) was made between the electrospray unit and the DMA. A further connection was made from the DMA to the CPC. This arrangement made access to these devices easier and allowed them to be changed as needed (Figure 2.28).

FIGURE 2.28 Components of commercial IVDS. Note MS2 peak on monitor.

The addition of a pressure vessel to hold the sample was a welcomed change. We struggled with capillary clogging in the original unit because it used a vacuum to pull the sample liquid from the vessel into the electrospray. This was not a correct method for handling samples for virus and nanoparticle detection. Pressure was used to push the sample through the capillary into the electrospray. We saw an immediate improvement in sample handling and the capillary functioned well afterward.

Software improvements helped processing of DMA and CDC data and we found it much easier to generate reports. The new IVDS unit proved a rugged and reliable instrument. It needed very little service over years of service despite the large volumes of samples analyzed.

2.5 RAPIDDX-IVDS

The current version of IVDS is morphing into a sleek new solid state device that takes the initial 1980s technology into the present. This device has incorporated advances. One major change is size. The RapidDx-IVDS is about a third of the size of the original IVDS. All the components easily fit into a small box. The RapidDx-IVDS has the capability of accommodating an autosampler to improve throughput and automate the analysis.

The upgrade to a CCD for counting the particles was a valuable addition. The CCD can count all the particles at the same time, thus eliminating the need for precise control to sweep through the particle sizes. The CCD makes detection time almost instantaneous. The sample needs a few moments to move from its container through the electrospray into the DMA, but it does not have to wait for several minutes while the instrument sweeps through the particle sizes.

Figure 2.29 shows some of these components. Notice the small size. Figure 2.30 is a diagram detailing the concept for the DMA-CCD operation.

An important feature of these upgrades is the handling of the data files along with the automation of the analysis and data stream. Results are stored in computer files and may be distributed electronically. This capability enables communications among multiple units and compilation analysis of data from several sources. It also allows analyzed data to be sent to command and control functions to aid decision

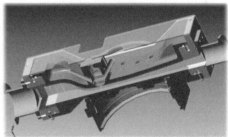

Photo: High Resolution DMA–Yale Univ.   CAD drawing shows internal flow-channels
       12-inch ruler for scale

FIGURE 2.29   Planar DMA components of RapidDx.

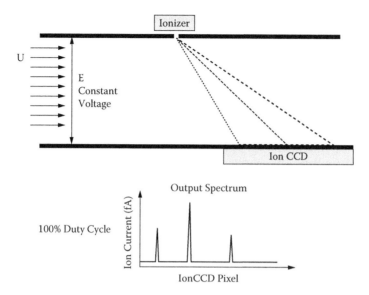

FIGURE 2.30 Concept of use with new CCD detector.

making. This is important because the automated operation allows continual monitoring of the normalized background. Such monitoring provides a basis for detecting the effects of other factors on an environment.

The normalized background can be expected to change with the time of day, activity in the area, seasons, and other common variables. If the normalized nanoparticle loading is known, any abrupt change can be considered an alert and trigger further investigation. If such information is distributed electronically to all relevant parties, the necessary response can be prompt and effective. Clearly, an alert for an unknown nanoparticle in the range of a known viral pathogen would merit attention. Networking these instruments provides useful spatial and temporal views of an environment.

2.6 CAPILLARY DIAMETER EFFECT AND IVDS STABILITY

2.6.1 INTRODUCTION

The IVDS instrument contains a capillary that transfers samples from vials into the detection section. The size of the capillary determines the amount of liquid that can be inserted into the instrument; the larger the capillary the more the volume. At times it may be useful to use a larger diameter capillary. The IVDS user should understand the relationship of capillary diameter, stability, and results obtained (Wick et al. 2010d).

This section will review the effect of changing the capillary inlet size of the electrospray injection module. Several inside diameter (i.d.) sizes of capillaries are available for the IVDS. The electrospray module aerosolizes the virus-containing solution and injects a monodispersed aerosol solution into the air stream for analysis.

The electrospray converts the sample to an aerosol by charging the liquid with an electric potential, pushing it through a capillary, and exerting an electric field at the capillary tip. The liquid evaporates from the droplets formed at the capillary tip and is carried into the sizing and counting modules of the IVDS. A change in the i.d. of the capillary may impact the detected concentration due to the close proximity of the capillary and the aerosolization of the virus sample.

The parameters associated with the electrospray module are air and CO_2 flow, sample overpressure, electrical voltage, and amperage. The air and CO_2 flow are fixed. The overpressure needs to be at a minimum (3 to 5 psi) to maintain a flow through the capillary. The only parameter allowed any variability is the setting of the voltage that exerts the electric field at the capillary tip. This parameter was controlled to a stable range of voltages when the MS2 bacteriophage was analyzed. Although the amperage changes with voltage, the amperage is not operator adjustable. As long as the voltages are within the stable range, the sample analyses will be consistent.

2.6.2 CAPILLARY PROCEDURES FOR MS2

Tests were run to determine the optimum i.d. for samples in the IVDS. The inside diameters of available capillaries were 25, 30, and 40 µm. The outside diameter (o.d.) did not change. A sample of MS2 was analyzed in the IVDS with the three available capillary sizes while other machine parameters were kept consistent. The scans were averaged for each size; the results are shown in Figures 2.31 through 2.33. The counts in the region of interest (ROI), from 22.5 to 28.9 nm, for MS2 increased with each increase of capillary i.d. Increases in MS2 counts were 3,206, 7,502, and 31,807 for the 25, 30, and 40 µm i.d. capillaries, respectively.

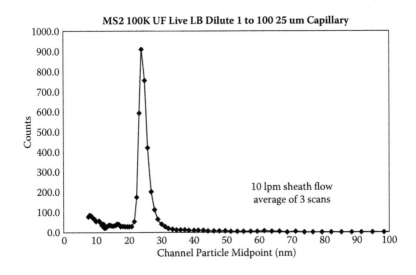

FIGURE 2.31   IVDS results for 25 µm inner diameter capillary.

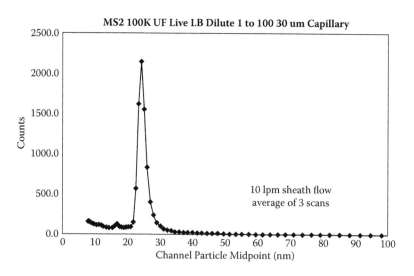

FIGURE 2.32   IVDS results for 30 μm inner diameter capillary.

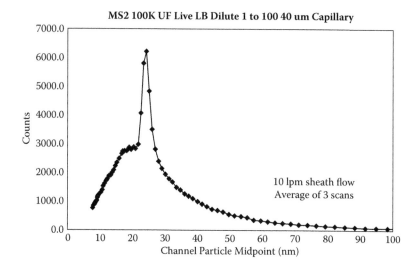

FIGURE 2.33   IVDS results for 40 μm inner diameter capillary.

2.6.3   CAPILLARY PROCEDURES FOR BUFFER SOLUTION

A similar set of analyses was performed on the standard 20 mM ammonium acetate buffer solution used in the IVDS. This buffer is used because of the very low
count rate across the IVDS scan area. The three capillary sizes were used to analyze
the ammonium acetate solution. The counts for the 25 and 30 μm capillaries were
acceptable across the scan range. The counts for the 40 μm capillary, although acceptable for many virus analyses, were much higher than the counts from the 25 and

30 µm capillaries. The higher count rates for the 40 µm capillary may interfere with analyses of smaller viruses in the IVDS.

2.6.4 RESULTS

The increase in counts from one capillary size to the next is 2.3 times for the 25 to 30 µm and 4.2 times for the 30 to 40 µm. However, the increase in i.d. area for the same capillaries is only 1.4 times for the 25 to 30 µm and 1.8 times for the 30 to 40 µm. The increase in counts may seem to be a positive factor in the analysis of viruses, i.e., more counts would make viruses easier to detect.

However, a closer analysis of the scans shows a marked increase in the background levels in the IVDS analysis using the 40 µm capillary. The background starts to reach an acceptable level only at ~100 nm. Most viruses have sizes up to ~150 nm and the increased background may hamper the analysis of many viruses. In addition, viruses smaller than 40 nm, such as the intestinal viruses between 25 and 35 nm, are of great interest due to the severity of illnesses they cause; they may be difficult to analyze at all due to the very high background counts seen in the largest capillary results.

The large background counts generated by the 40 mm capillary would interfere with viral analysis unless the sample virus concentrations were very high. The high background would also make any attempt at sample quantification virtually impossible. The recommended capillary sizes for virus analysis with the IVDS are the 25 and 30 µm sizes. However, we strongly recommend knowing the capillary in use at the time of analysis and not mixing capillaries when performing similar analyses on samples. Mixing capillaries, especially during counting or concentration studies could lead to misleading results. The 40 mm capillary may be indicated where high flow rates and increased volumes are indicated.

2.7 ELECTROSPRAY STABILITY

We tested the stability of the electrospray injection module of the IVDS system. This module aerosolizes the virus-containing solution and injects a monodispersed aerosol solution into the air stream for analysis. It converts the sample to an aerosol by charging the liquid with an electric potential, pushing it through a capillary, and exerting an electric field at the capillary tip. The liquid evaporates from the droplets formed at the capillary tip and is carried into the sizing and counting modules of the IVDS.

The parameters associated with the electrospray module are air and CO_2 flow, sample overpressure, electrical voltage, and amperage. The air and CO_2 flows are not variable. The overpressure needs to be at a minimum (between 3 and 5 psi) to maintain flow through the capillary. The only parameter with any variability is the setting of the electrical voltage exerting the electrical field at the capillary tip. This parameter was varied across the stable range of voltages, and the MS2 bacteriophage was analyzed. Although the amperage changes with voltage, the amperage is not operator adjustable.

The electrical voltage at the capillary tip that changes the shape of the liquid flow from the tip was thought to need optimization for repeatable IVDS analysis. This study explores the variations in electrospray voltage and the effects on IVDS performance.

2.7.1 Procedure

A sample of MS2 stock solution (0.5 mL) was diluted with 50 mL of distilled water and filtered with 50 mL of 20 mM ammonium acetate buffer solution by the ultrafiltration (UF) subsystem of the IVDS using a 100 KDa filter. The filtration system was intended to remove all materials with molecular weights smaller than the level for which the filtration system is set (100 KDa in this case), e.g., growth media, salt molecules, and proteins from the solution, and leave a concentrated virus solution. The filtered MS2 solution was then diluted at 1:10 with the ammonium acetate buffer to increase its conductivity to allow it to be injected into the IVDS.

The sample of MS2 was analyzed by adjusting the electrospray voltage from the minimum that produced a relatively unstable cone to the highest voltage that produced a relatively stable cone. A stable cone would not exhibit any large fluctuations in shape or cause pulsations that would disrupt the process. Figure 2.34 shows the range of minimum, optimum, and maximum electrospray cones observed. The electrospray cone produces large drops instead of a stable cone at the minimum range.

The electrospray voltage was ramped in 0.05 kV increments, and three IVDS scans were acquired and saved for each voltage. The amperages were recorded, and

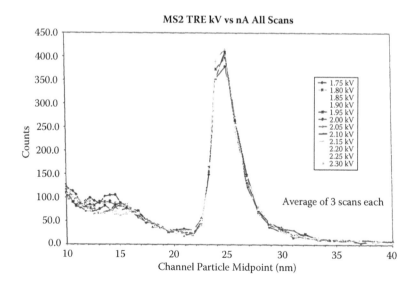

FIGURE 2.34 Range of minimum, optimum, and maximum electrospray cones.

the scans were output to MS Excel for analysis. The region of interest (ROI) for MS2 is 22.5 to 27.9 nm. The counts over the ROI were totaled and tabulated for analysis.

2.7.2 RESULTS

Table 2.3 lists the kV, nA, sum of ROI, and average and standard deviations of the counts. The counts averaged 1438 ± 28 for 12 sets of kV scans and were in very close agreement over the large range of applied voltages. Plotting the kV versus nA in Figure 2.35 shows the linearity (R2 = 0.9942) of the stable portion of the electrospray cone. The amperage tracks closely with the voltage. Figures 2.36 through 2.38 show the output scans generated by the IVDS.

Each voltage scanned represents an average of three individual scans. The scans are in very close proximity to each other, as verified by the ROI counts obtained from the numerical data files. The change in peak counts was consistent even when the voltages changed significantly. A wide range of voltage conditions can be used to produce consistent results. The output count data from the IVDS for this MS2 sample was very consistent even with the large variations in applied voltage. The electrospray cone in a stable configuration will yield repeatable results from the IVDS. Combining electrospray stability with the ability to use different capillary configurations gave a wide latitude to IVDS operations. This flexibility could be helpful for optimizing procedures in the future.

TABLE 2.3
Electrospray Stability Study: kV, nA, Sum of ROI, and Average and Standard Deviations of Counts

kV	nA	Sum of ROI Counts (22.5–27.9 nm)		
1.75	−228	1457.4		
1.80	−232	1458.5	Count average	
1.85	−235	1454.9		1438
1.90	−238	1408.9	Standard Deviation	
1.95	−243	1446.1		28
2.00	−249	1430.9		
2.05	−254	1437.6		
2.10	−260	1411.9		
2.15	−264	1392.1		
2.20	−269	1404.9		
2.25	−274	1465.4		
2.30	−280	1483.3		

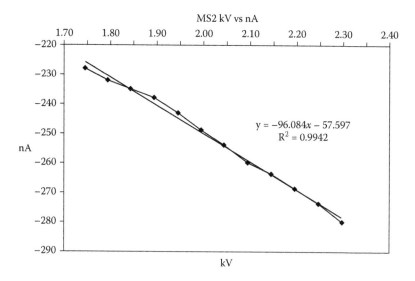

FIGURE 2.35 Linearity (R2 = 0.9942) of stable portion of electrospray cone.

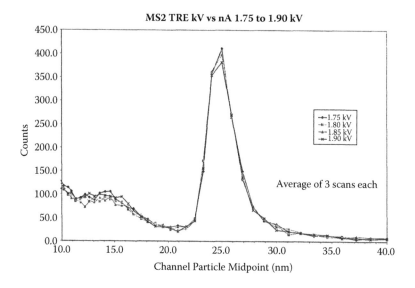

FIGURE 2.36 IVDS results for electrospray stability—1.75 to 1.90 kV.

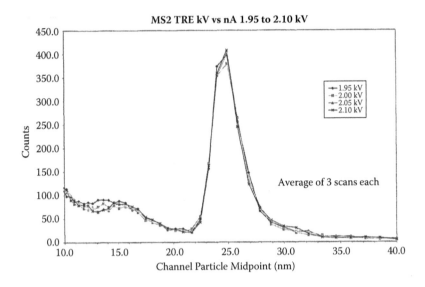

FIGURE 2.37   IVDS results for electrospray stability—1.95 to 2.10 kV.

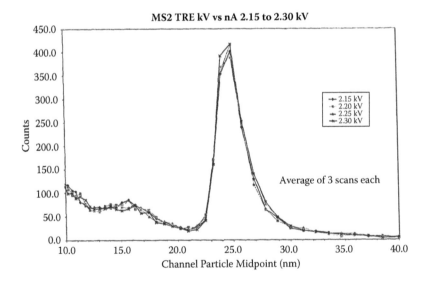

FIGURE 2.38   IVDS results for electrospray stability—2.15 to 2.30 kV.

3 Proof Concept Using MS2

3.1 INTRODUCTION

This chapter puts together many aspects of virus detection and the use of IVDS that demonstrate this technology. The physical basis of the invention that makes this work will be discussed. Two important virus features enable this technology (1) viruses have different sizes, and (2) they have different physical properties that set them aside from other nanoparticles. These two distinct attributes will be discussed.

It was important to select a test virus. It had to be one that was easy to grow and safe for users. MS2 was selected and will be discussed. Likewise it was important to use this test virus to evaluate the new technology. Several examples are given. Other viruses and their results are presented in Chapter 7.

3.2 PHYSICAL BASIS OF INVENTION

3.2.1 DIFFERENT SIZED VIRUSES

Viruses are nanoparticles. They are an order of magnitude smaller than bacteria that in turn are generally an order of magnitude smaller than fungi (Wick 2014). We will consider the uniqueness of the viruses as we consider how to detect and identify them.

Figure 3.1 shows the approximate ranges of particles reaching nanometer size. Although several particles would appear to overlap the viruses, they have very different physical properties from viruses. Figure 3.2 shows the different sizes for various viruses. This chart is important for a couple of reasons. We have determined that some of these viruses, for example MS2, are excellent nanometer standards. They are stable and remarkable in that they are always the same size and can be used for years without issues.

For some, their peaks are more defined than polystyrene beads. The more polymorphic viruses add a bit of a challenge because their presented optical size covers a wider range than the tightly packed RNA environmental viruses. Viruses such as influenza are represented on an IVDS chart as a peak at either 92 or 102 nm (depending on the strain of influenza) with a wide base. This is because of the appendages on the virus and its elongated shape. It should be noted that this is the only virus that causes as fuzzy line on a density gradient for the same reasons. Nevertheless, it is consistent and always follows the physical properties associated with it. This consistency is expected to be likewise with other viruses.

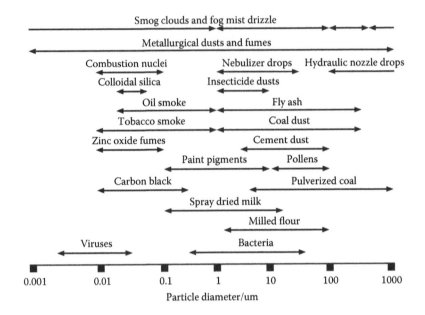

FIGURE 3.1   Different sized nanoparticles.

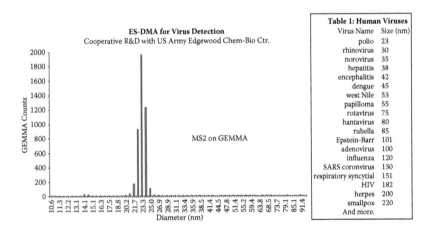

FIGURE 3.2   Different sized viruses. Note IVDS output with detection peak for MS2.

IVDS was not able to test some of the more odious viruses simply because of the hazards they present. It is recommended that these viruses be examined by the appropriate experts to gain experience with how they would present themselves in an IVDS analysis. The 4 nm resolution of the current IVDS allows for many viruses to be separated from each other. For example, there are many sizes of hepatitis viruses, similar to respiratory viruses and many others. Improvements are expected to improve this resolution to less than 1 nm and this may provide some separation even among viruses that are currently very close in size. It has been demonstrated, as another example, that different loading of a viral coat can change the size of a virus by just a little, but the change is discernible with the current IVDS.

The nominal sizes of human virus species have sufficient disparity in size so that the IVDS method can distinguish one from another. IVDS data is displayed as a histogram showing the size distributions of virus particles in a sample. If a particular nanoparticle-size virus consistently appears at high concentration in samples from at-risk populations, the sharp peak signals a new infection, even in advance of disease symptoms. Confirmatory genetic measurements by polymerase chain reaction (PCR) and other tests may be necessary for a definitive identification, but as a rapid virus screening technique capable of running large numbers of samples at points of care, the portable IVDS method is very powerful and unique.

Indeed, real-time knowledge that a virus is present in a patient may be useful in its own right even in advance of any definitive identification. Automation such as that provided by the RapidDx will make it possible to generate large numbers of spectra for virus analysis data from a wide variety of samples. To the extent possible, these data can be collected online or transmitted to others to collate a library of results. Statistical methods can be used to filter this data to seek additional information and determine trends in order to generate detection algorithms that will be programmed back into the RapidDx for the rapid detection of viruses. This technology may enable generation of a vast data base of virus information leading to rapid virus triage. Size does make a difference when discussing virus detection.

3.2.2 VIRUS WINDOW

The virus window (Figure 3.3) was described in Chapters 1 and 2 in some detail. For completeness, a figure and discussion are included here because of the importance of this feature of viruses. As noted earlier, viruses exhibit different properties from nanoparticles of similar size and thus exist in a specific physical environment that facilitates their separation by physical means from various matrices. As discussed in Chapter 1 and expanded upon in Chapter 2, it is possible to separate the various virus families from each other simply because of their different sizes as shown in Section 3.2.1.

In addition to the virus size differences and the virus window, other issues should be discussed. The ability to achieve a 3D address for each virus group or family makes it possible to automate the collection and identification of these viruses as discussed in Chapter 1. Furthermore, the families that overlap each other are frequently not found in the same samples. The concentrations of viruses in humans, insects, plants, and other animals differ throughout the environment. A virus causing a problem such as an infection can be expected to have a higher concentration than a virus that does not cause infections. This difference became abundantly clear after we examined more than 30 viruses associated with honeybees (see Chapter 10).

It is not even necessary to know which family of viruses is being detected in most operational cases. The presence of a higher-than-normal count of any virus is an item of interest. This ability to determine baseline viral loading and be able to monitor it over a long period is a useful view. A temporal view of viruses gives an opportunity to use them as secondary measures of what is taking place in their environments.

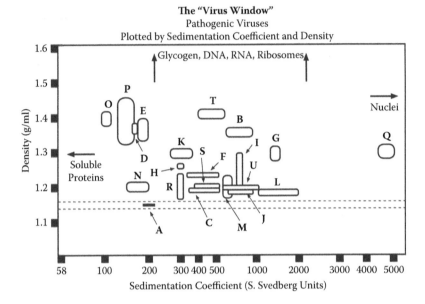

FIGURE 3.3 Virus window. Key: A: Microsomal background, B: Adenoviridae, C: Arenaviridae, D: Astroviridae, E: Calicivirdae, F: Coronaviridae, G: Filoviridae, H: Hepadnaviridae, I: Herpersviridae, J: Orthomyxviridae, K: Papovaviridae, L: Paramyxviridae, M: Retroviridae, N: Flavivaridae, O: Parvovaridae, Q: Poxviridae, R: Togaviridae, S: Bunyaviridae, T: Reoviridae, and U: Rhabodoviridae.

These uses for virus detection and classification are outside the normal use of the virus window as it was first envisioned. The window was used to monitor bacterial loading in seawater by monitoring virus loading including the bacteriophages associated with various bacteria. It is easier to collect and determine the virus load of water than it is to classify all the bacteria in the same sample. This concept was applied to the study of honeybees discussed in Chapter 10. Cause-and-effect curves could be generated for various treatments of the honeybees while monitoring of rises and declines in virus loading continued. This type of monitoring study is very useful for managing honeybee health.

When considering the other information in the virus window, it becomes clear that knowing that some virus families are widely separated from each other in their three-dimensional (3D) addresses is useful when collecting temporal data. Counting peaks will appear at different places and are monitored easily. In cases where the families are closer to each other, the 4-nm resolution of the IVDS instrument becomes useful and the temporal peaks become relevant for many viruses.

The use of the information in the virus window is not restricted to these few examples. The physical features of viruses allow them to be further examined. Unknown viruses—created by humans or emerging from the environment—are just as likely to follow the physical behavior patterns of their relatives. An unknown peak in a temporal virus loading chart invites further collection and examination. Due to the ease of collecting, analyzing, and plotting different sized viruses, it makes sense to use such information to monitor environments, people, plants, and animals for unusual changes in their natural conditions.

3.3 MS2 AS TEST VIRUS

The MS2 bacteriophage was first used as an example virus particle (Wick et al. 1999a; Wick et al. 1999b; Wick et al. 2005a). Its use is standard practice in testing detection systems because it is environmentally stable and not infectious. MS2 is grown along with its host bacteria via standard procedures along with its host bacteria. Other test viruses were obtained from other sources such as animal tissues and fluids and tissue culture media.

3.4 CHARACTERIZATION OF MS2 USING IVDS

3.4.1 INTRODUCTION

The ability to detect and analyze viruses has been a goal of science for many years, since the discovery that a new type of microorganism was responsible for diseases in both humans and animals. Viruses are smaller than bacteria and their size presented the first challenge. Their small size made their classification difficult and the field of virology advanced through biochemical techniques rather than by direct examination. Recent advances in electron microscopy helped the detection and classification issues and much information has been reported on the physical features of many virus families. These historic techniques are, however, time consuming and require special knowledge, and specialized reagents.

We learned that it was possible, based on the physical characteristics of viruses, to count individual viruses directly in a new and dramatic way using the IVDS instrument described in Chapters 1 and 2. The IVDS uses readily available materials and is simple to operate.

This new IVDS capability allowed us to seek a reliable, useful, and easily available virus particle that could be used for calibration testing. The virus particle had to be noninfective and easy to grow and store. IVDS was used to analyze and characterize a sample of the MS2 bacteriophage provided by the Life Sciences Division at Dugway Proving Ground (DPG). This highly purified sample consisted of 2 mL of purified MS2 bacteriophage at a concentration 1×10^{14} pfu/mL or 10.2 mg protein/mL from Lot #98110. The MS2 sample was analyzed using the IVDS instrument. The GEMMA detector represented one stage of the technique. An electrospray unit was used to inject samples into the detector. Other components were a differential mobility analyzer and a condensate particles counter.

3.4.2 RESULTS

A high-purity MS2 sample with 1×10^{14} pfu/mL (hereafter described as DPM14) was analyzed using the GEMMA virus detector. The sample was placed neat into the GEMMA and the results are shown in Figure 3.4. The graph shows a very high virus count (over 150,000) and other features. MS2 is nominally 24 to 26 nm in size, as illustrated in Figure 3.4. In fact, the sample as received was difficult to aspirate through the GEMMA capillary delivery system because of the high concentration.

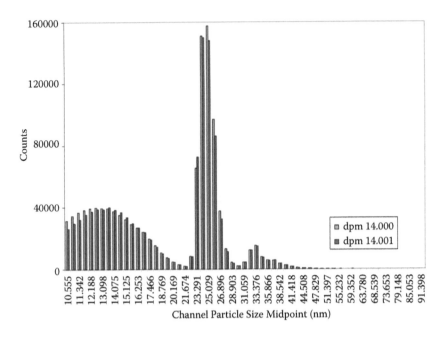

FIGURE 3.4   MS2, 1 × 1014 pfu/mL, DPG Lot #98110.

When the difficulty of sampling the neat DPM14 MS2 sample became apparent, the sample was then serially diluted to produce a number of lower concentration samples. An aliquot of DPM14 was diluted tenfold to produce a sample of MS2 at a concentration of 1×10^{13} pfu/mL and designated DPM13. The dilutions were all made with a 0.02 M solution of ammonium acetate (pH ~10) required for the electrospray unit. The pH was adjusted to keep the virus from breaking down into its component subunits. Sample DPM13 was then diluted tenfold as were the following dilutions. Table 3.1 lists the samples produced by serial dilutions of the original sample.

Figures 3.5 through 3.10 show the effects of serial dilution on the MS2 sample. One observation is the accuracy of the IVDS in showing the effects of the dilutions.

TABLE 3.1
IVDS Physical Counts for MS2 Samples

	Counts in Size Range				
MS2 Sample	25.946 nm	25.029 nm	24.144 nm	23.291 nm	22.468 nm
DPM8			1		
DPM9		2	5	3	
DPM10		17	88	52	
DPM11		146	929	541	78
DPM12	148	3613	12582	5174	255
DPM13	15216	57624	65021	16893	1664
DPM14	96995	157461	150886	65389	8347

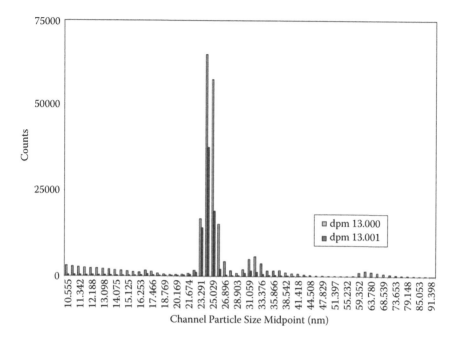

FIGURE 3.5   MS2, 1×10^{13} pfu/mL, DPG Lot #98110.

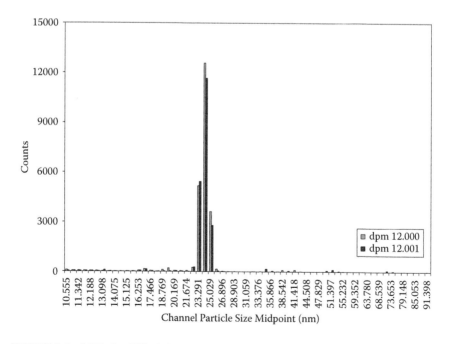

FIGURE 3.6   MS2, 1×10^{12} pfu/mL, DPG Lot #98110.

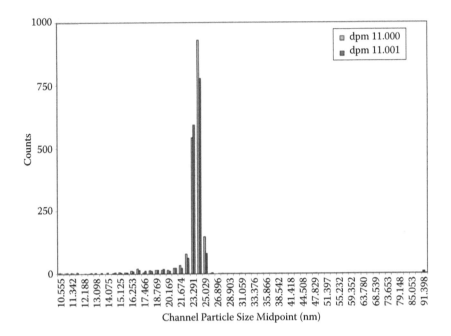

FIGURE 3.7   MS2, 1×10^{11} pfu/mL, DPG Lot #98110.

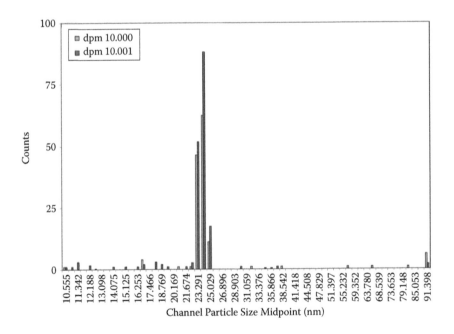

FIGURE 3.8   MS2, 1×10^{10} pfu/mL, DPG Lot #98110.

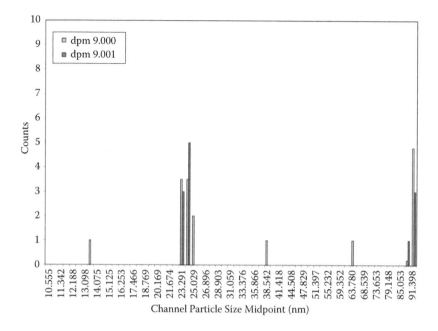

FIGURE 3.9 MS2, 1×10^9 pfu/mL, DPG Lot #98110.

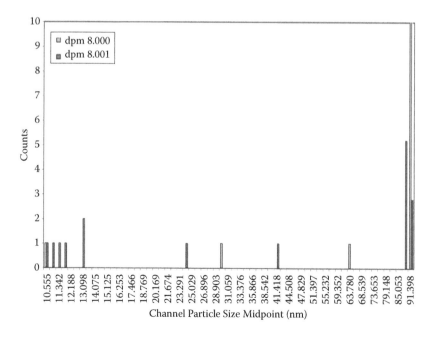

FIGURE 3.10 MS2, 1×10^8 pfu/mL, DPG Lot #98110.

The lowest dilution (DPM8) shows the results of filtration and the absence of other nanoparticles. No attempt was made to improve the counting sensitivity in these experiments; the intent was to use MS2 as a suitable test virus.

3.4.3 ANALYSIS AND DISCUSSION

The GEMMA unit easily detected the MS2 bacteriophage. The virus is consistently detected in the range of 22 to 26 nm. The GEMMA scans also show very low backgrounds away from the MS2 peaks. The action of serially diluting the MS2 did not affect the stability of the bacteriophage in solution. In fact, the additions of ammonium acetate buffer to produce dilutions reduced the background counts. The GEMMA scans of buffer solutions show very low count because ammonium acetate is nearly invisible to the detector.

The count rates for the various concentrations of MS2 are tabulated in Table 3.2. A comparison of the multiplication factor from sample to sample was also tabulated. The lower concentrations display a fairly consistent multiplier and are consistent with the target dilutions. As the concentrations increase, the multiplier appears to decrease in magnitude. As noted in Section 1.1, the as-received DP14 sample was difficult to aspirate into the GEMMA detector. This sample was very concentrated and this appeared to interfere with the analysis.

The reduction in the multiplier may have been caused by the agglomeration of particles as they flowed through the CPC of the GEMMA unit. This agglomeration would lower the number of particles counted and reduce the multiplier. It would appear that a count rate over 100,000 in a few adjacent channels with a virus in this size range (~25 nm) approaches the upper limit to concentrations that can be analyzed in the detector. This is remedied by simply diluting a sample to less than 100,000 counts in adjacent channels.

The actual sensitivity of the GEMMA detector was not in question in this study. The solution presented to the detector could have been concentrated further to allow analysis of samples that appeared too dilute. The DPM8 sample could be concentrated from its original 1 mL volume to 10 µL. This would then present the GEMMA

TABLE 3.2
Numerical Analysis of MS2 Peak Count Data

MS2 Sample	Sum of Size Range	Sample-to-Sample Multiplier
DPM8	1	—
DPM9	10	10.0
DPM10	157	15.7
DPM11	1694	10.8
DPM12	21772	12.9
DPM13	156418	7.2
DPM14	479078	3.1

with a sample that would generate a graph with ~100 counts in a scan. The number of viruses that can be detected by the GEMMA is very low, on the order of 10. Thus its ability to detect viruses is only a function of the concentration of the presented solution.

In a simple experiment, a few thousand viruses were measured into 500 mL of water. The water sample was concentrated through the ultrafiltration unit and nearly 800 viruses were counted by the GEMMA. The limiting factor for analysis is the ability to further concentrate a liquid solution and effectively handle it without losing it due the handling problems arising from tiny volumes.

The MS2 bacteriophage sample was very pure and concentrated. No other viruses were detected. The sample responded well to serial dilutions and was stable in the ammonium acetate buffer. This technique is a simple method to test the purity of any virus preparation since the IVDS instrument is not limited to detection of specific viruses.

3.5   REMOVING COMPLEX GROWTH MEDIA

This section discuses purification of the MS2 bacteriophage from complex growth media and the resulting analysis by the IVDS.

3.5.1   INTRODUCTION

Virus detection and analysis present many inherent challenges. One of the more important is the process of purifying and concentrating a virus from its background material. This is required for all detection methods be used in subsequent steps. The background loading that may contain growth media, salts, proteins, and other materials present purification and concentration challenges and there is little point to even considering virus detection until the background issue is resolved and samples are purified of growth media, salts, proteins, and other impurities (Wick et al. 1999b).

A 500-mL sample of grown MS2 bacteriophage with growth media was received from the Life Sciences Division at Dugway Proving Ground (DPG). The virus concentration was 1.4×10^{12} pfu/mL. The growth media was composed of L-B broth, 10g tryptone, 10 parts NaCl, and 5 parts yeast extract. The solution was dark yellow and clear.

The MS2 sample was analyzed using the IVDS-Beta instrument connected to a GEMMA detector and an ultrafiltration module. The GEMMA detector consisted of an electrospray unit to inject samples into the detector, a DMA and a CPC.

Several solutions were prepared to explore the ability of the ultrafiltration apparatus to remove contaminants and retain viruses of interest in solution. A sample of albumin from chicken eggs was prepared at a concentration of 0.02%, by weight in an ammonium acetate (0.02 M) buffer. MS2 bacteriophage to a concentration of 3×10^{11} pfu/mL was added to the solution. A second solution containing 2.5% cesium chloride (CsCl), by weight was also prepared in the same buffer and MS2 bacteriophage to a concentration of 5×10^{11} pfu/mL was added. Both MS2 bacteriophage samples were highly purified.

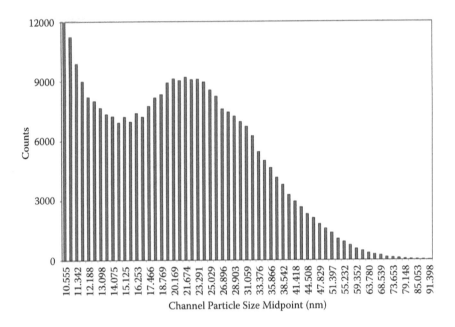

FIGURE 3.11   GEMMA analysis of MS2 bacteriophage plus growth media, DPG Lot #98251.

3.5.2   RESULTS OF MS2 WITH GROWTH MEDIA

The mixed MS2 sample with 1.4×10^{12} pfu/mL was analyzed using the GEMMA detector. The sample was placed neat into the detector and the results are shown in Figure 3.11. The growth media with the MS2 in solution produces a graph displaying very broad, nondescript peak across the area of interest of 24 to 26 nm—the expected size for the MS2 bacteriophage. It is not apparent from the as-received sample analysis whether the sample actually contained MS2 in solution. The solution required removal of the growth media before meaningful results could be obtained.

3.5.2.1   Ultrafiltration of MS2 with Growth Media

The virus sample including growth media was purified and concentrated using an ultrafiltration (UF) process to concentrate virus species for further study. The UF stage is a hollow fiber-based tangential or cross-flow filtration system that operates by pumping a feed stream through the hollow fiber. As the solution passes through the fiber, the sweeping action of the flow helps prevent clogging of the fiber. A pressure differential forces the filtrate through the fiber, while the virus feed stream is purified and concentrated. A wide range of fiber pore sizes are available. This filtration technique can reduce volumes from over 5 to 0.2 mL. A sample of the DPG MS2 with growth media was then processed through the ultrafiltration apparatus. The parameters for ultrafiltration are listed in Table 3.3.

TABLE 3.3

Ultrafiltration Parameters

for MS2 in Growth Media

Initial sample volume	3 mL
Pump speed	2
Transducer pressure	15 psig
Total buffer wash volume	50 mL
Final sample volume	2 mL
MWCO of module	500K

By continually washing the sample volume with ammonium acetate buffer (GEMMA working fluid), the UF will remove ions, proteins, and all other materials smaller than the 500K molecular weight cutoff (MWCO) of the cross-flow filter. The bacteriophage will be retained in the circulating solution and continue to be purified by the process. Because the 500K MWCO filter will effectively retain the MS2, the total wash volume can be significantly larger than the initial sample volume. The ultrafiltration of the sample was completed in less than 10 minutes.

3.5.2.2 Counting MS2 after Removal of Growth Media

After ultrafiltration, the sample was then analyzed by the GEMMA. The results are shown in Figure 3.12. The graph shows that most, if not all, of the growth media was

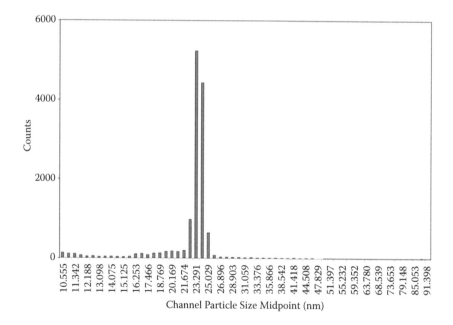

FIGURE 3.12 GEMMA analysis of MS2 (growth media sample), ultrafiltration processed.

TABLE 3.4

GEMMA Counts for MS2 Bacteriophage

Channel Midpoint Diameter (nm)	Counts	Channel Midpoint Diameter (nm)	Counts
10.5545	128.2	32.1968	17.5
10.9411	105.7	33.3762	8.5
11.3419	97.7	34.5989	10.5
11.7574	64.3	35.8664	7.5
12.1881	37.3	37.1803	6.5
12.6346	50.8	38.5423	2.5
13.0975	34.2	39.9542	3
13.5773	36.8	41.4178	6.8
14.0746	39.5	42.9351	2.2
14.5902	34.6	44.5079	5
15.1247	28	46.1384	2
15.6788	41.5	47.8286	1
16.2531	102.1	49.5807	2
16.8485	110.8	51.397	1
17.4658	80.3	53.2798	0
18.1056	120.7	55.2316	1
18.7688	129.4	57.2549	0
19.4564	167	59.3523	0
20.1691	175.2	61.5265	1
20.908	168.6	63.7804	1
21.6739	192.2	66.1169	0
22.4679	973.1	68.539	1
23.291	5228.2	71.0497	1
24.1442	4429.7	73.6525	1
25.0287	639.9	76.3506	0
25.9455	73.6	79.1476	1
26.896	31.4	82.047	0
27.8813	25.6	85.0526	0
28.9026	22	88.1683	2
29.9614	19.5	91.3982	0
31.059	16.2		

removed and replaced with the ammonium acetate buffer which is virtually invisible to the GEMMA. The remaining solution from the ultrafiltration apparatus was very pure and satisfactory for further experimentation. The numerical results in Table 3.4 show a very low count rate, essentially a background level, outside the area of MS2 peaks. At this level of purification and concentration, the original 500 mL of as-received MS2 with growth media could yield more than 300 mL of very pure highly concentrated MS2 bacteriophage in ammonium acetate.

3.5.3 RESULTS OF MS2 WITH ALBUMIN

A sample of 0.02% albumin in ammonium acetate to which was added 3×10^{11} pfu/mL of MS2 bacteriophage was analyzed neat in the GEMMA. As shown in Figure 3.13, the MS2 peak is centered around 24 nm. The albumin in the sample is displayed as a very broad peak starting below 10 nm and extending to 20 nm.

A sample of albumin plus MS2 was then processed through the ultrafiltration apparatus. The parameters for ultrafiltration are shown in Table 3.5. After ultrafiltration, the sample was examined in the GEMMA. As shown in Figure 3.14, the only

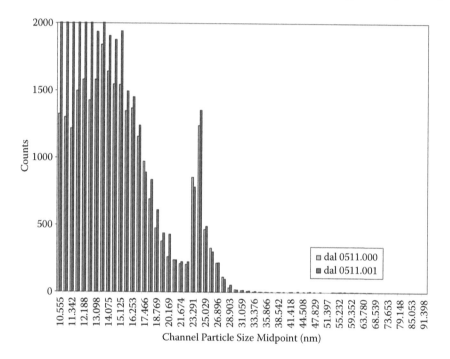

FIGURE 3.13   GEMMA analysis of albumin in ammonium acetate with MS2 (3×10^{11} pfu/mL).

TABLE 3.5
Ultrafiltration Parameters
for Albumin plus MS2

Initial sample volume	1 mL
Pump speed	2
Transducer pressure	15 psig
Total buffer wash volume	40 mL
Final sample volume	0.4 mL
MWCO of module	500K

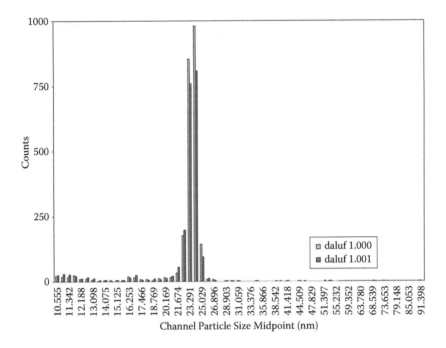

FIGURE 3.14 GEMMA analysis of albumin plus MS2, ultrafiltration processed from Figure 3.13.

peak in evidence is centered at 24 nm. The large peak between 10 and 20 nm was removed. The processing of the sample through the ultrafiltration apparatus completely removed the albumin protein and retained the MS2 bacteriophage.

3.5.4   RESULTS OF MS2 WITH CESIUM CHLORIDE

A sample of 2.5% CsCl by weight in ammonium acetate to which was added 5×10^{11} pfu/mL of MS2 bacteriophage was analyzed neat in the GEMMA. As shown in Figure 3.15, the MS2 peak is centered around 24 nm. The CsCl in the sample is displayed as a very broad peak starting below 10 nm and extending to over 20 nm. Any higher concentration of CsCl would start to obscure the MS2 peak position.

3.5.4.1   Ultrafiltration of MS2 with Cesium Chloride

The sample of CsCl plus MS2 was then processed through the ultrafiltration apparatus. The parameters for ultrafiltration are shown in Table 3.6.

3.5.4.2   Counting MS2 after Removal of Cesium Chloride

After ultrafiltration, the sample was examined in the GEMMA. As shown in Figure 3.16, the MS2 peak is centered on 24 nm. Most of the large peak between 10 and 22 nm was removed. A small remnant of the CsCl peak in the processed sample resulted from the smaller amount of buffer wash volume in this cycle. To completely remove the CsCl, the ultrafiltration process would need to be continued with further

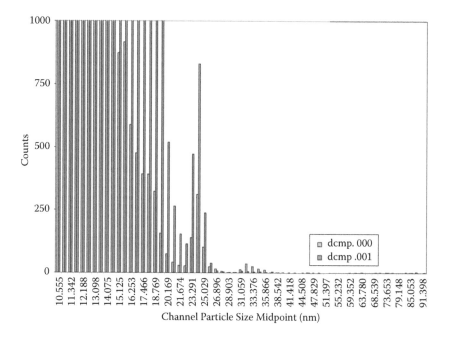

FIGURE 3.15   GEMMA analysis of cesium chloride (2.5%) plus MS2 (5×10^{11} pfu/mL).

TABLE 3.6
Ultrafiltration Parameters
for CsCl plus MS2

Initial sample volume	1 mL
Pump speed	2
Transducer pressure	15 psig
Total buffer wash volume	30 mL
Final sample volume	0.5 mL
MWCO of module	500K

washing until all the salt was replaced with buffer solution. The processing of this sample through the ultrafiltration apparatus also retained the MS2 bacteriophage.

3.5.5   MS2 Purification

The ultrafiltration apparatus effectively removed the growth media from the various solutions of MS2 bacteriophage. The addition of approximately ten times the amount of starting solution with ammonium acetate buffer (3 mL versus 50 mL, respectively) allowed the efficient replacement of growth media with the buffer solution. The background of the GEMMA scan of the ultrafiltration-processed solution was very low due to the low detection of ammonium acetate. The ultrafiltration for comparable volumes can be completed in ~10 minutes.

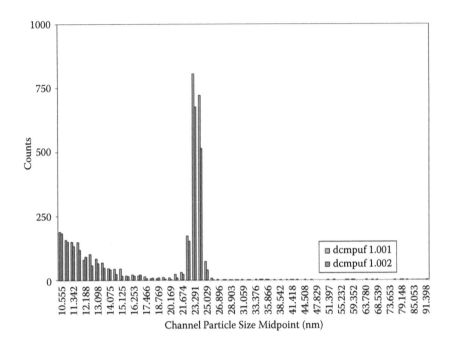

FIGURE 3.16   GEMMA analysis of cesium chloride plus MS2 with ultrafiltration.

Other contaminating materials present in a virus solution can also be removed successfully from solution while retaining the virus. The albumin protein was removed almost completely from the MS2-containing solution by ultrafiltration. Pore size adjustment of the ultrafiltration module provides great flexibility in the processing of solutions.

The CsCl solution appeared to require further washing to completely remove the salt from the virus-containing solution. The tests to date indicate that the wash volume for the removal of CsCl in the ultrafiltration apparatus requires the initial sample volume to be washed with about 40 to 50 times the volume of buffer solution to achieve complete removal of certain impurities.

3.5.6   Use of MS2 for Testing: Conclusions

The sample of MS2 bacteriophage containing original growth media was purified by ultrafiltration. The growth media were removed and the MS2 retained. The ultrafiltration module was equipped with a 500 K molecular weight cutoff cross-flow filter that effectively retained the bacteriophage while allowing removal of the growth media. The remaining solution of bacteriophage was ready for further characterization and purification. This technique should be effective for removing similar materials from solutions of other viruses.

Samples of laboratory-produced solutions, specifically albumin protein and CsCl solutions with MS2 bacteriophage were processed successfully by the ultrafiltration apparatus that removed these typical contaminants and retained the bacteriophage.

3.6   FILTRATION CHARACTERISTICS OF MS2 BACTERIOPHAGE

3.6.1   INTRODUCTION

This study of MS2 and filter performance resulted from my discovery that the new IVDS instrument did not achieve the level of MS2 detection expected because the particles were flowing past the filter used for sample preparation. This analysis of filters allowed us to resolve the IVDS detection problem and improve its performance (Wick et al. 1999c).

3.6.2   FILTER STUDY

The MS2 bacteriophage has a reported nominal molecular weight of 2 MDa and would not be expected to pass through a filters of various sizes with low molecular weight cutoff (MWCO) values below 1 MDa. We found that MS2 will pass through filters with 750, 500, and 300 KDa values. MS2 was retained on the 100 KDa filter. A cross-flow hollow fiber apparatus was used for the 750 and 500 KDa analyses. Centrifuge filters of 1 MDa and 300 and 100 KDa were tested. The rate of passage of MS2 through the cross-flow filters is dependent on tangential flow rate and pressure. Passage through the centrifuge filters depended on the gravitational force applied.

Nominal MWCO values of various filters can lead to the assumption that items larger than the cutoff values will be retained after filtration. We found, at least for the MS2 bacteriophage, that there are exceptions. During the operation of the IVDS instrument, counts of MS2 decreased during ultrafiltration and purification. It was further observed that this did not occur when the MS2 passed through a filter and was not retained as expected. This was an important discovery because even the loss of a small number of viruses limits sensitivity. These observations led to further investigations.

To better understand the filtration characteristics of the MS2 bacteriophage, we explored the abilities of various filtration techniques and their relative effectiveness. A purified sample of MS2 bacteriophage was used in the filtration studies.

The two types of filters used in this study were a centrifuge tube assembly (the solution is forced through the filter by gravitational forces) and a cross-flow filter apparatus (pressure pushes the solution through the filter). The centrifuge filter assemblies are available in various sizes and MWCO filter inserts. The MWCO may be changed to capture biological materials such as proteins, cell products, and viruses via molecular weight differentiation. The cross-flow filter or ultrafiltration apparatus can also capture or reject biological materials by adjustment of the MWCO of the filter.

These filtration systems operate by pumping a feed stream through a hollow fiber. As the solution passes through the fiber, the sweeping action of the flow helps prevent clogging of the fiber. A pressure differential forces the filtrate through the fiber, while the biological feed stream is purified and concentrated. A wide range of pore sizes for centrifuge and hollow fiber filters are available.

The MS2 samples were analyzed after filtration using the GEMMA detector component of the IVDS. The GEMMA consists of an electrospray unit to inject samples into the detector, a differential mobility analyzer (DMA) and a condensate particles counter (CPC). Chapters 1 and 2 provide complete descriptions of all the components of the IVDS.

3.6.2.1 Slant (Wedge) Filters

The first set of solutions consisted of 1×10^{11} pfu/mL of MS2 in a cesium chloride (CsCl) solution (0.5%, by weight) in an ammonium acetate buffer (0.02M). The procedure was to place 150 μL of the solution into a wedge filter of differing MWCO values (30, 50, and 100 KDa). The filter was then centrifuged and the samples were analyzed in the GEMMA. As shown in Table 3.7, the wedge filter concentrated the MS2 solution, i.e., the counts increased as the solution size decreased. Even with a subsequent addition of buffer and recentrifugation, the solutions continued to concentrate.

The same solution (CsCl 0.5% + 1×10^{11} pfu/mL MS2) was then placed into a 1 MDa centrifuge filter and spun. The first concentration shows an increase (Table 3.7) from 150 to 350 counts in the sample. The solution volume decrease from 1,000 to 100 μL was expected to increase the counts measured. The subsequent wash and recentrifugation should have shown a further increase in MS2 counts. However, the counts for the washed sample were even lower. The conclusion from the filtration with the 1 M MWCO filter is that the MS2 bacteriophage can pass through the filter and is not retained.

The filtrate was analyzed to determine whether the MS2 passed through the centrifuge filters. A separate sample of 1×10^{12} pfu/mL MS2 (DPG, ultrafiltration cleaned, mixed media sample) was filtered with the 1 M centrifuge filter. As shown in Figure 3.17, the MS2 passed through the filter and was deposited in the filtrate. Table 3.8 presents the numerical counts from the GEMMA analysis of the retentate after one wash cycle and the filtrate from the 1 M centrifugation of the sample.

To determine whether the CsCl caused interference during the filtration with the 1 M filter, a solution of MS2 was prepared at a concentration of 1×10^{11} pfu/mL by dilution in the ammonium acetate buffer only. The solution was then centrifuged in the 1M filter. As shown in Table 3.9, the plain MS2 solution also passed through the 1M filter apparatus with a loss of virus material. The CsCl did not appear to affect the loss of virus material by its presence in the filtration solution.

TABLE 3.7
Filtration of MS2 plus CsCl Solutions

Sample	Filter MWCO (Da)	Counts	Volume (μL)	+1 Wash (Count)	Volume (μL)
CsCl 0.5% + 1×10^{11} MS2, DPG	None	150	150		
CsCl 0.5% + 1×10^{11} MS2, DPG	30K	2500	25	4500	35
CsCl 0.5% + 1×10^{11} MS2, DPG	50K	2000	20	3000	25
CsCl 0.5% + 1×10^{11} MS2, DPG	100K	9000	15	5000	10 (+5 buffer)
CsCl 0.5% + 1×10^{11} MS2, DPG	1 M centrifuge	350	100	75	50

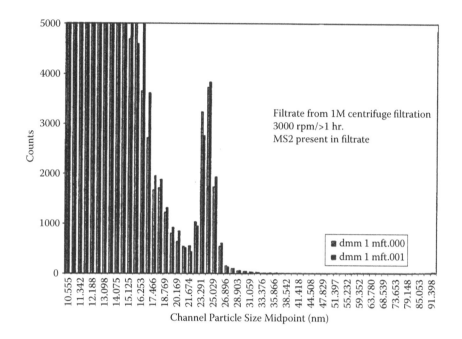

FIGURE 3.17   Filtrate from 1M (MWCO) centrifugation of MS2 solution.

TABLE 3.8
Filtration of MS2 Solution after Ultrafiltration

Sample	Filter MWCO (Da)	GEMMA Count	Volume (µL)	+ 1 Wash (Count)	Volume (µL)
DPG MS2 Mixed Media UF Mod 1	None	5,000	100		
DPG MS2 Mixed Media UF Mod 1 Retentate	1 M centrifuge			75	100
DPG MS2 Mixed Media UF Mod 1 Filtrate	1 M centrifuge	3,500	150		

TABLE 3.9
Filtration of Pure MS2 Solutions

Sample	Filter MWCO (Da)	GEMMA Count	Volume (µL)
1×10^{11} MS2, DPG	None	600	100
1×10^{11} MS2, DPG Retentate	1 M centrifuge	65	100

3.6.2.2 Cross-Flow or Tangential Flow Filters

Another type of filtration is the cross-flow or tangential flow technique. A solution to be examined is pumped through a hollow fiber designed to allow the passage of differing MWCO materials, depending on the filter installed. A flow restriction at the exit from the fiber bundle develops a pressure differential that forces the filtrate through the fiber and concentrates the feed solution.

The first sample prepared for filtration was a CsCl (0.05%, by weight) solution with 3×10^{11} pfu/mL MS2 added into an ammonium acetate buffer. The ultrafiltration parameters for this solution are shown in Table 3.10. As shown in Table 3.11, the sample volume was concentrated from 1,000 to 100 µL and the MS2 counts dropped from 3,200 to 25. The count drop indicates that the cross-flow filter (MWCO of 750 KDa) allowed the virus to pass through the hollow fiber.

The second sample, a CsCl solution (2.5%, by weight) plus 5×10^{11} pfu/mL MS2 in ammonium acetate buffer, was processed through the cross-flow filtration apparatus with a 500 K MWCO value. The parameters for the ultrafiltration processing of the solution are shown in Table 3.12. Although the sample volume was concentrated by

TABLE 3.10
Cross-Flow Parameters for CsCl (0.05%) plus MS2 (3×10^{11} pfu)

Sample volume-initial	1 mL
Pump speed	2
Transducer pressure	15 psig
Total buffer wash volume	40 mL
Sample volume-final	0.1 mL
MWCO of module	750K

TABLE 3.11
Cross-Flow Filtration of CsCl (0.05%) plus MS2 (3×10^{11} pfu)

Sample	Filter MWCO (Daltons)	GEMMA Counts	Volume (µL)
CsCl 0.05% + 3×10^{11} MS2, DPG	None	3200	1000
CsCl 0.05% + 3×10^{11} MS2, DPG Retentate	UF Mod 750K	25	100

TABLE 3.12
Cross-Flow Parameters for CsCl (2.5%) plus MS2 (5×10^{11} pfu)

Sample	Filter MWCO (Da)	GEMMA Counts	Volume (µL)
CsCl 2.5% + 5×10^{11} MS2, DPG	None	800	1000
CsCl 2.5% + 5×10^{11} MS2, DPG Retentate	UF Mod 1500K	750	500

TABLE 3.13

Cross-Flow Filtration for CsCl (2.5%) plus MS2 (5 × 10¹¹ pfu)

Sample	Filter MWCO (Da)	GEMMA Counts	Volume (µL)
CsCl 2.5% + 5 × 10¹¹ MS2, DPG	None	800	1000
CsCl 2.5% + 5 × 10¹¹ MS2, DPG Retentate	UF Mod 1500K	750	500

half, the counts remained constant, as shown in Table 3.13. The MS2 virus also passed through the 500 K filter, although at a slower rate than shown by the 750 K filter.

To test the lower limit of MWCO for a MS2 bacteriophage, we used a centrifuge filter of 300 K MWCO. It appears from Table 3.7 that filters up to 100 K MWCO do not allow the passage of MS2 through a filter medium. The 300 K filter was loaded with 100 µL, diluted to 1 mL in ammonium acetate buffer, of 1×10^{11} pfu/mL MS2. The sample was centrifuged and the retentate analyzed. As shown in Figure 3.18, the MS2 is at least partially retained in the 300 K filter.

To determine the amount, if any, of MS2 passing through the filter, a 1 mL portion of the filtrate was concentrated in 100 K wedge filters. The final volume was reduced to 25 µL. As shown in Figure 3.19, MS2 was present in the filtrate from the 300 K centrifuge filtration. It appears that the MS2 is able to pass through MWCO filters as small as 300 K; it does not appear to pass through the 100 K centrifuge filters.

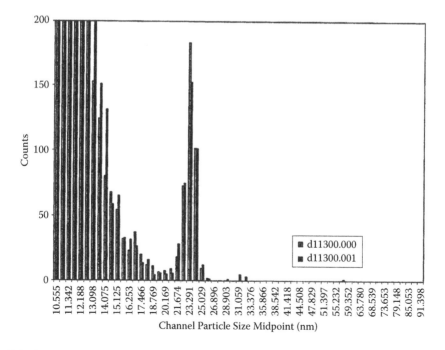

FIGURE 3.18   GEMMA scan of 1×10^{11} pfu MS2 retained on 300K MWCO centrifuge filter.

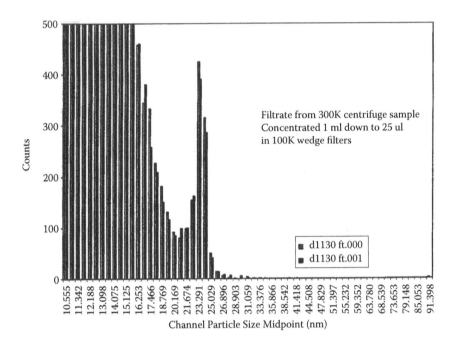

FIGURE 3.19 GEMMA scan of 1×10^{11} pfu MS2 concentrated filtrate.

A series of solutions of 1×10^{12} pfu/mL of MS2 bacteriophage was then filtered by the cross-flow apparatus with a 750 K MWCO ultrafilter. All the filtered solutions included 1 mL of the 1×10^{12} pfu/mL MS2 with various additions of ammonium acetate buffer solution. The additions of buffer allowed differing lengths of time of filtration in the cross-flow apparatus, while keeping the amount of MS2 in the sample constant. However, the concentration of the MS2 will vary based on the dilution factor in the starting sample.

The samples were processed in the cross-flow apparatus until concentrated to ~1 mL volume of the 1×10^{12} pfu/mL MS2 initial sample. Table 3.14 presents the filtration

TABLE 3.14

Cross-Flow Parameters for MS2 (1×10^{12} pfu) with Variable Volume Ammonium Acetate Buffer Solutions

Initial sample volume	1 mL MS2 + variable buffer volumes
Pump speed	2
Transducer pressure	15 psig
Total buffer wash volume	Variable
Final sample volume	0.70 to 0.75 mL
MWCO of module	750K

TABLE 3.15

Dilution Amounts and GEMMA Analysis of Cross-Flow Filtration of MS2 Samples

MS2 Start Volume (pfu/m)	Ammonium Acetate Dilution (mL)	Final Volume (mL)	GEMMA Counts for MS2 Peaks (two-run average)
1 mL @ 1 × 10^{12}	0	1.0	9255
1 mL @ 1 × 10^{12}	1	0.75	5164
1 mL @ 1 × 10^{12}	2	0.70	5280
1 mL @ 1 × 10^{12}	4	0.75	3239
1 mL @ 1 × 10^{12}	8	0.70	5284
1 mL @ 1 × 10^{12}	16	0.75	3549
1 mL @ 1 × 10^{12}	32	0.70	2830

parameters for the cross-flow apparatus for this set of experiments. Table 3.15 shows the starting volumes, initial dilutions, final sample volumes, and subsequent GEMMA sample counts for the MS2 viral peaks.

The final volume of the solutions processed through the cross-flow apparatus was essentially equivalent. The solutions should therefore exhibit the same count rate for MS2 because the initial amount of virus was equal in all cases. The count rates are plotted in Figure 3.20, and show a logarithmic decline as the dilutions were increased. The increased dilution time lengthened the contact time with the cross-flow filter and subsequently increased the loss of the MS2 bacteriophage through the filter medium.

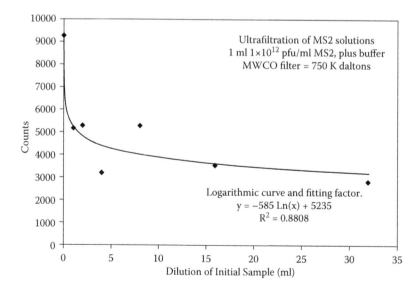

FIGURE 3.20 GEMMA counts and logarithmic curve for variable dilutions and cross flow. Filtration of MS2 plus ammonium acetate buffer solutions.

3.6.3   RESULTS OF FILTER STUDY

The MS2 bacteriophage was able to pass through the filters with MWCO values of 300 KDa and larger and was retained on filters of 100 K and smaller. This result was not expected because MS2 has an approximate size of 2 MDa and was expected to be retained on the initial size-tested 750 K filter. This confirmed observations and counts plotted showing a logarithmic decline as dilutions were increased (Figure 3.20). This study showed the retention of MS2 with MWCO filters of 100 KDa and smaller and its passage through a 500 KDa filter.

The variable dilution cross-flow filtration analysis shows the logarithmic removal of the MS2 from the feed stream as the solutions were concentrated. The longer the MS2 solution was in contact with the cross-flow filter of 750 K, the more MS2 was removed from the solution. If the goal of cross-flow filtration is to remove salts and other ionic entities, a smaller MWCO filter (such as 100 K) should be used and the MS2 would be retained.

However, to remove larger macromolecules from a sample of MS2 bacteriophage, a different approach is needed. A larger MWCO filter (macromolecule-dependent) should be used to retain and concentrate the macromolecule while the MS2 bacterio-phage is removed in the filtrate stream. The filtrate stream could then be processed sep-arately with a 100 K MWCO filter to retain and concentrate the MS2 bacteriophage. The extra step would only add a short time to an analysis because cross-flow filtration is a fast and efficient step. Filtration is discussed in other chapters because it is a critical step in separating, concentrating, and otherwise preparing samples for analysis.

3.7   SUMMARY

- The MS2 bacteriophage passed through various MWCO filters larger than 300 K and was retained on the 100 K and smaller filters.
- Variable dilutions with cross-flow filtration apparatus and a 750 K MWCO filter yielded a logarithmic removal of MS2.
- Retention of MS2 on 300 K-rated hollow fibers was high enough that only minor losses were incurred under conditions that would accomplish signifi-cant removal of even large proteins.
- We clearly need a better understanding of molecular weight cutoff (MWCO) and more investigation is needed to examine pore size impacts.
- This finding has important implications for investigations and procedures dependent upon retaining viruses.

4 Concentration of Viruses

4.1 INTRODUCTION

A concentrator device and method of concentrating a liquid sample were invented (Wick 2012a). A concentrator device may include a pressure vessel and a filter element disposed within the pressure vessel. The pressure vessel may include an inlet configured to introduce pressurized air into a first portion of the pressure vessel and a first outlet fluidly coupled with a second portion of the pressure vessel. The first outlet may be adapted to be opened and closed selectively. A second outlet may be configured to receive a capillary tube inserted into the first portion of the pressure vessel.

The filter element may be configured to receive a liquid sample to be concentrated. The element may separate the first portion of the pressure vessel from the second portion substantially and may define a retentate side adjacent to the first portion and a filtrate side adjacent to the second portion. When pressurized air is introduced through the inlet and the first outlet is open, a filtrate of the liquid sample received in the filter element may pass from the retentate side to the filtrate side such that a concentrated retentate of the liquid sample remains on the retentate side. When the first outlet is closed, the concentrated retentate of the liquid sample may be forced out through the capillary tube (Wick 2012a).

4.2 CONCENTRATOR DEVICE

4.2.1 BACKGROUND

The concentrator device can relate to devices, systems, and/or methods for concentrating a liquid sample and also to the detection, identification, and monitoring of submicron-sized particles in concentrated samples such as, for example, viruses, virus-like agents, prions, viral subunits, and viral cores of dilapidated viruses.

Detection and identification of viruses pathogenic to humans in an environment without limiting the detection and identification to a particular family, genus, and species can be difficult, particularly in field conditions such as a potential biological warfare (BW) threat environment or an environment containing an emerging new microbe. Instruments are needed to allow detection of, for example, remote releases of biological agents in a field environment, thereby providing early warning capabilities and allowing calculations for troop movements and wind patterns.

In the detection and monitoring of viruses, false positives associated with background materials can be major obstacles. Background materials may include biological or other debris that obscures virus detection by being registered as viruses when a sample is analyzed. Analysis of viruses requires a high degree of purification of the viruses to overcome background loading and avoid false positives. For example, a

BW virus may be buried within loadings of other microorganisms that form biological debris having loading on a magnitude of 10^{10} larger than the threshold loading for the targeted virus.

Methods that culture viruses can often be used to increase virus load over background material. Culture methods, however, may be too slow for effective BW detection. Furthermore, some important viruses cannot be cultured easily. Viruses may also be extracted from an environment and concentrated to an extent that permits their detection and monitoring without culturing procedures.

Generally, in the detection of small amounts of viruses in environmental or biological liquids, it is necessary to both enrich the concentration of viruses many orders of magnitude (greatly reduce volume of liquid solubilizing the viruses) and remove non-viral impurities. In the presence of non-viral impurities, even the most sensitive detection methods generally require virus concentrations on the order of 10 femtomoles per microliter or more in the sampled liquid to reliably detect viruses.

In general, a standard method for the concentration of virus samples involves a tangential (cross) flow filtration system to reduce the volume of the sample while removing impurities such as salts and small cellular debris. A more practical and effective device is needed, for example, for use with a capillary inlet of a gas phase electrophoretic mobility molecular analysis (GEMMA) device.

4.2.2 NEW CONCENTRATOR DEVICE

In an embodiment of the invention, a liquid sample concentrator device may include a pressure vessel and a filter element disposed within the pressure vessel. The pressure vessel may include an inlet configured to introduce pressurized air into a first portion of the pressure vessel and a first outlet fluidly coupled with a second portion of the vessel. The first outlet may be adapted to be opened and closed selectively. A second outlet may be configured to receive a capillary tube inserted into the first portion of the vessel.

In an embodiment of the invention, a method for concentrating a liquid sample may be provided. The method may include introducing a predetermined amount of a liquid sample to be concentrated into a filter element disposed within a pressure vessel. The method may include introducing pressurized air into a first portion of a pressure vessel on a retentate side of the filter element and opening a first outlet fluidly coupled with a second portion of the pressure vessel whereby a filtrate of the liquid sample passes through the filter element to the second portion from the retentate side. The method may include closing the first outlet when a predetermined amount of the liquid sample remains on the retentate side of the filter or after a predetermined amount of time. When the first outlet is closed, a concentrated retentate of the liquid sample is forced through an inlet end of a capillary tube positioned proximate to the retentate side of the filter element.

4.2.3 COMPONENTS AND FUNCTIONS

An embodiment of the invention is discussed in detail below. Specific terminology is employed for the sake of clarity. However, the invention is not intended to be limited to the specific terminology so selected, and it should be understood that the terminology is used for illustration purposes only. A person skilled in the relevant art will recognize that other components and configurations can be used without parting from the spirit and scope of the invention. Each specific element includes all technical equivalents that operate in a similar manner to accomplish a similar purpose.

In the following description of an embodiment of the invention, directional words such as *top, bottom, left, right, upwardly, downwardly, clockwise,* and *counterclockwise,* are employed by way of description and not limitation with respect to the orientation of the device and its various components as illustrated in the drawings.

Figure 4.1 depicts a concentrator device (module) 10 constructed to concentrate a liquid sample S according to an embodiment of the invention. Filtered concentration of a liquid sample (e.g., volume reduction coupled with retention of specific submicron-sized particles) may allow for more effective detection, identification, and monitoring of submicron-sized particles in the sample S.

As shown in the figure, the concentrator device 10 may include a pressure vessel 12 having a first portion 14 and a second portion 16. An inlet 18 may be configured to introduce pressurized air into the first portion 14 of the pressure vessel 12. A first outlet 20 may be coupled fluidly with the second portion 16 of the pressure vessel 12 and may be adapted to be opened and closed selectively, for example, with a valve mechanism 22.

A second outlet 24 may be configured to receive a capillary tube 28 inserted into the first portion 14 of the pressure vessel 12. The capillary tube 28 may be coupled to an electrospray unit of a GEMMA device (not shown) as described in Chapter 1, Figure 1.2.

A filter element 26 may be disposed within the pressure vessel 12 and configured to receive a liquid sample S to be concentrated. The filter element 26 may substantially separate the first and second portions 14, 16 of the pressure vessel 12 and may define a retentate side adjacent to the first portion 14 and a filtrate side adjacent to the second portion 16. An electrically conductive wire 30, for example, a platinum wire, may be coupled to a voltage source of the electrospray unit of the GEMMA device

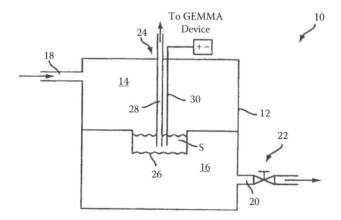

FIGURE 4.1   Device for concentrating liquid sample.

and extend through the first portion 14 into the retentate side of the filter element 26 to establish a current return for the electrospray operation.

The liquid sample S may have an initial volume, for example, of approximately 500 µL when received in the filter element within the pressure vessel. When pressurized air is introduced through the inlet 18 and the valve mechanism 22 at the first outlet 20 is open, a filtrate of the liquid sample S may pass from the retentate side to the filtrate side of the filter element 26 such that a concentrated retentate of the liquid sample S remains on the retentate side.

The pressurized air is not limited to clean, regulated air supplied under approximately 3 to 4 psi; other pressures may be sufficient, depending on the properties of the various components of the device, particularly filter element 26. When the concentrated retentate remaining on the retentate side of the filter element 26 reaches a predetermined volume, e.g., approximately 80 to 100 µL or alternatively a predetermined amount of time passes, the valve mechanism 22 of the first outlet 20 may be closed (manually by a user or automatically by a control device).

When the valve mechanism 22 of the first outlet 20 is closed, the remaining concentrated retentate of the liquid sample S may be forced out through the capillary tube 28 onto the electrospray unit of the GEMMA device. In this way, the concentrator device 10 may utilize an existing overpressure at the capillary inlet of the electrospray unit to filter and concentrate a virus sample. This would allow analysis by the GEMMA without removal of the concentrated sample from the IVDS.

Figure 4.2 depicts an exploded view of the concentrator device 100. It may include a pressure vessel 102 (see Figure 4.3) having a hollow outer housing 104 and an inner (base) member 106 constructed to be received in an open end of the hollow outer housing 104. The pressure vessel including the housing and inner member may be formed from any suitable material, for example, a thermoplastic material such as poly(methyl methacrylate (PMMA), glass, metal, or a combination thereof.

The concentrator device may also include a filter element for receiving a liquid sample to be concentrated that may be utilized in the pressure vessel. Because a virus-containing sample does not come into contact with the pressure chamber, the material used to make the pressure chamber is not critical. Clear plastic allows an operator to see the level of the sample. This is useful because as the volume decreases the operator can turn off the device and send samples to the GEMMA at various intervals.

The hollow outer housing 104 may be a substantially cylindrical container although other configurations and shapes are possible. The housing may have an open end and an inner surface 114 defining interior volume. The housing 104 may include an inlet 112, a first outlet 122, and a second outlet 116. A valve mechanism 124 and 126 may be coupled to the first outlet 122 and may be any suitable pressure release device including a thumb valve or similar assembly. The valve mechanism may include an outlet 124 and a flow control element 126 that may toggle between open and closed positions; it may be operated manually or automatically by a control device. The second outlet 116 may be configured to hold a threaded fitting 118 for receipt of a capillary tube from the electrospray unit An electrically conductive wire 120 may extend from the fitting 118 into the interior volume of the housing 104.

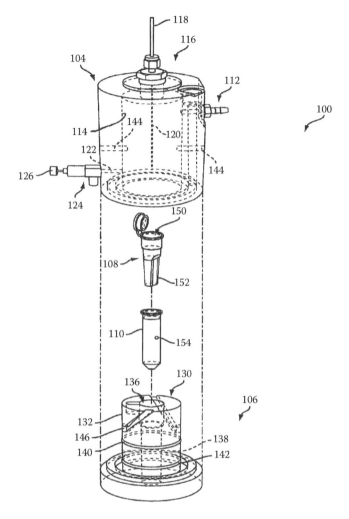

FIGURE 4.2 Exploded view of concentrator.

The inner member 106 may include a base portion and an upper portion 130 configured to be received through the open end of the hollow outer housing 104. Like the housing, the inner member may be substantially cylindrical. The upper portion of the inner member may include an outer surface 132 and define an interior cavity 136 constructed to receive the filter element 108 when the concentrator device 100 is assembled. A passageway 138 between the interior cavity and the outer surface 132 is provided. The upper portion 130 of the inner member may also include one or more seal elements 140, 142 such as, for example, polymeric or rubber O-rings, disposed about the outer surface 132 at positions both above and below the passageway 138.

When the concentrator device 100 is assembled and the inner member 106 is received in the housing 104, the interior cavity 136 passageway 138 may be coupled fluidly with the first outlet 122 by way of an annular volume defined (radially) between

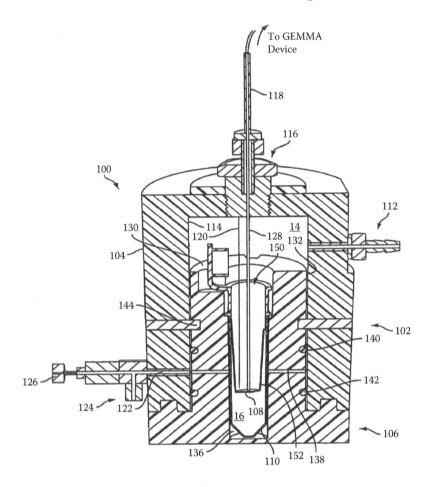

FIGURE 4.3   Cross-sectional view of assembled concentrator shown in Figure 4.2.

the outer surface 132 and inner surface 114 as well as (longitudinally) between seal elements 140, 142. When assembled (see Figure 4.3), the inner member 106 and the hollow outer housing 104 may be coupled removably to one another. In the embodiment depicted in Figures 4.2 and 4.3, a bayonet connection is shown wherein one or more pegs 144 may be disposed on the inner surface 114 of the housing 104 for receipt in one or more angled slots 146 defined on the outer surface 132 of the upper portion 130.

The filter element 108 may be a centrifugal ultrafiltration wedge filter such as a Ultrafree® 0.5 centrifugal filter unit available from Millipore Corporation of Billerica, Massachusetts. The filter has a predetermined molecular weight cutoff value (MWCO) of 100 KDa. The filter element may include a filter membrane of polyethersulfone, for example, and may define a retentate side 150 and a filtrate side 152. The filter element may optionally be received in a filtrate receiver 110 having an opening 154 configured to allow fluid (or pressure) communication between the filtrate side of the filter element and the interior cavity of the inner member.

Figure 4.3 depicts a cross-sectional view of the assembled concentrator device 100 shown in Figure 4.2. The inner member 106 may be received in the hollow outer housing 104 to define the pressure vessel 102. The filter element containing a liquid sample S to be concentrated may be received in the interior cavity 136 of the inner member 106. The pressure vessel 102 is then coupled to the electrospray unit of the GEMMA device via the capillary tube 128 from the electrospray unit that extends through the fitting 118 at the second outlet 116 and into the liquid sample on, the retentate side of the filter element 108.

The electrically conductive wire extends into the liquid sample on the retentate side of the filter element to establish a current return for the electrospray operation. Clean, regulated air pressured at 3 to 4 psi is then introduced through the inlet 112. The first outlet 122 may be opened by opening the valve mechanism 124, 126. A filtrate of the liquid sample S disposed in the filter element 108 may pass from the retentate side 150 to the filtrate side 152 so that a concentrated retentate of the liquid sample S remains on the retentate side. When a predetermined amount of the liquid sample S remains on the retentate side of the filter element or after a predetermined time, the first outlet may be closed by closing the valve mechanism 124, 126, and the concentrated retentate of the liquid sample S is then forced out through the capillary tube 128.

The concentrator device 100 may be utilized as a modular handheld device or may be incorporated into an IVDS. The device may eliminate a time-consuming centrifugation step that may take 60 minutes or more. The concentrator may allow direct analysis of the liquid sample after concentrating. Tests were performed to determine concentration efficiency with a virus sample as shown in the following examples.

4.2.4 TESTING

Figure 4.4 depicts GEMMA results for an initial MS2 bacteriophage sample without concentration of the sample in the device shown in Figures 4.2 and 4.3 (in the capillary inlet sample holder). A 500-μL sample of MS2 bacteriophage (initial concentration of ~1 × 10^5 particles per milliliter) was introduced by pipetting into the sample holder and the sample was analyzed by the GEMMA. The results of the count in the region of interest (ROI) are shown in Figure 4.4. An initial count in the ROI was 531, measured between 23.3 and 27.9 nm.

Figure 4.5 is a depiction of the GEMMA results for a sample of MS2 after a 4-minute concentration in the concentrator device. A 500-μL sample of initial concentration of ~1 × 10^5 particles per milliliter was pipette into a centrifugal ultrafiltration wedge filter. The filter was then inserted into the concentrator module and the sample was concentrated for 4 minutes to 100 μL and analyzed with the GEMMA device. The results of the count in the ROI are shown in Figure 4.5. The initial count in the ROI (region of interest) was 8674, measured between 23.3 and 27.9 nm—a 16-fold increase in initial counts in ROI over the unconcentrated sample. In addition, processing in the concentrator removed the large salt peak below 13 nm.

Figure 4.6 illustrates GEMMA results for an MS2 bacteriophage sample based on measurements of partial concentrations taken in 1-minute increments in the

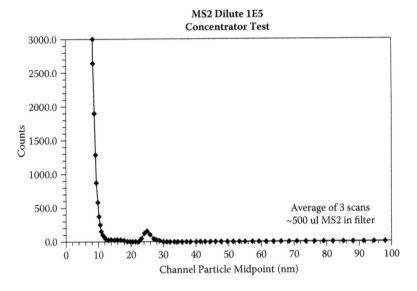

FIGURE 4.4 IVDS results for initial MS2 bacteriophage sample.

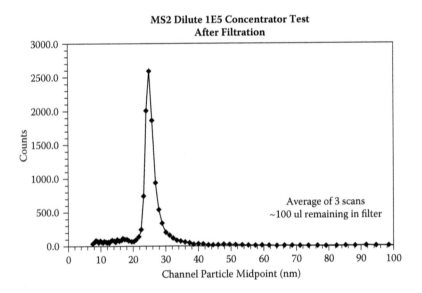

FIGURE 4.5 IVDS results for MS2 bacteriophage sample after 4 minutes in concentrator device shown in Figures 4.2 and 4.3.

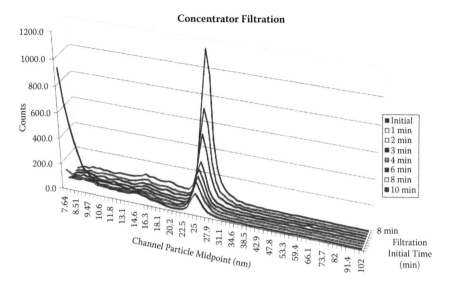

FIGURE 4.6 IVDS results for MS2 bacteriophage sample based on measurements of partial concentrations taken in 1-minute increments in concentrator.

concentrator. A 500-µL sample at an initial concentration of approximately 1×10^5 particles per milliliter was pipetted into a centrifugal ultrafiltration wedge filter. The filter was then inserted into the concentrator module. The sample was concentrated and filtered in 1-minute increments and analyzed at the end of each increment by the GEMMA. The filtration was stopped when the sample reached a volume of approximately 80 µL at 10 minutes.

The results of the count at the end of each 1-minute increment in the ROI are shown in Figure 4.6. The MS2 counts in the ROI (23.3 to 27.9 nm) increased with each increment of filtration and are listed in Table 4.1. As show in Figure 4.6,

TABLE 4.1

MS2 Counts Taken at 1-Minute Increments as Shown in Figure 4.6

MS2 Counts from Timed Concentrator Analyses

Time (min)	ROI Counts (23.3 to 27.9 nm)
0	559
1	812
2	883
3	1023
4	1153
6	1793
8	2432
10	3744

concentration and filtration of the sample in the concentrator removed the large salt peak below 13 nm after the first minute of filtration and the subsequent scans were clean below 15 nm. It should be noted that the concentrator can be used to concentrate particles in several ways. In summary, the concentrator provides a means to reduce the volume of a sample and thus increases the number of viruses per unit of volume and the number of particles counted by the GEMMA. The concentrator

- Is portable or can be integrated into IVDS
- Can concentrate sample volumes from about 1 mL to 0.1 mL
- Is capable of sending samples to GEMMA in increments
- Allows hands-free operation

5 Cause and Effect Curves

5.1 INTRODUCTION

Cause and effect curves or dose response curves are very helpful in determining how to treat medical problems. In the classic sense, these curves have been used to test the effectiveness or efficacy levels of antibiotics.

A regimen of antibiotics is tested against a bacterial infection. Periodic sampling then reveals the bacterial counts over time and a chart is developed. Effective treatments reduce the bacteria counts and less effective treatments do not. Various adjustments to doses of an antibiotic can be tested and a proper dose response curve can be developed. The curves are useful for determining the most useful dose for treating an infection.

Likewise, honeybees can be treated with a fungicide to control fungal pests. The proper dose would remove the fungi but not kill the honeybees. The measure of effectiveness is improved honeybee health. In this case, the dose response curve was developed using honeybee viral loading as the measure of health. Reduced viral loading was equated to improved honeybee health.

Such curves can be extended to determine environmental conditions such as temperature and chemical treatments such as changes in pH. The effectiveness of sunlight and other treatments for biological decontamination can be analyzed and a cause and effect curve developed. This chapter contains examples showing effects of temperature and pH levels on the survival of the MS2 virus.

5.2 EFFECTS OF TEMPERATURE ON MS2 SURVIVAL

The stability of viruses under different environmental conditions had always been a problem for microbiologists. Measuring this stability requires subjecting a virus to a harsh environment and monitoring the decay of the number concentration of virus particles over time. Until recently, measuring stability was exceedingly difficult. However, using IVDS has simplified this approach since the device can characterize and measure the concentrations of viruses (Wick et al. 2005a).

5.2.1 PROCEDURES

A sample of MS2 stock solution (2 mL) was obtained from the National Institute of Standards and Technology (NIST). The stock solution was diluted with 100 mL of distilled water and filtered by the ultrafiltration (UF) subsystem of the IVDS using 100 KDa filters. The function of the filtration system is to remove all materials with molecular weights smaller than the filtration system set for (100 KDa in this case) such as growth media, salt molecules, and proteins from the solution and leave a concentrated virus solution.

The concentrated MS2 solution was added to about 20 mL of 20 mM ammonium acetate solution. The ammonium acetate increases the conductivity of the solution and allows it to be injected into the test module of the IVDS. The MS2 solution was concentrated again by the UF subsystem to a total of 2.5 mL of clean solution. The bacteriophage solution was then subjected to elevated temperatures and high and low pH values. The IVDS periodically extracted and analyzed the samples to determine the dose response curves for temperature and pH. This same method can be followed for treatment with antiviral agents or other factors to determine the response to a target virus over time, then build a dose response curve.

5.2.2 RESULTS

The results of analysis of the original MS2 stock solution (after purification and concentration) are shown in Figure 5.1 and Figure 5.2. Figure 5.1 shows the counts of the MS2 at around 24 nm. Figure 5.2, an expanded scale of Figure 5.1, shows a peak at around 14 nm that represents a breakdown protein of the virus.

Figure 5.3 shows the number of virus particles counted as a function of temperature. The residence time at each temperature was 10 minutes. We can see that the number of virus particles declined somewhat (by about 25%) at temperatures of 62 to 63°C after 10 minutes. Further increase of temperature to 64°C caused a sharp decline in the number of surviving virus particles; the number of surviving particles continued to decrease as the temperature rose. The unexpected increase in the number of particles at 70°C is probably due to disassociation or clumping of particles as the virons disintegrate.

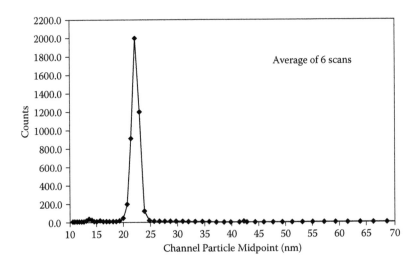

FIGURE 5.1   MS2 initial sample (six-scan average).

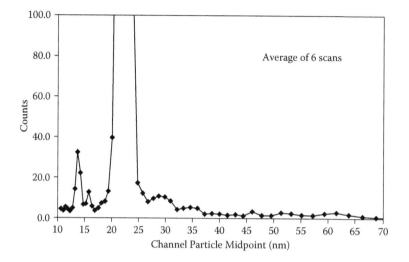

FIGURE 5.2   MS2 initial sample showing fragments at −14 nm (six-scan average).

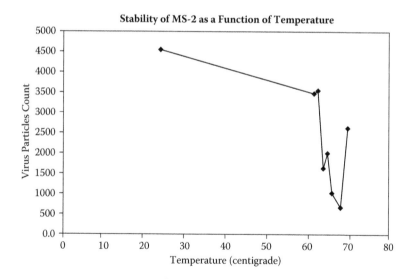

FIGURE 5.3   Stability of MS2 as function of temperature.

Figure 5.4 shows the number of particles counted as a function of time when the MS2 stock solution was exposed to 63°C. Some drop in the number of virus particles occurred in the first 30 minutes. In the next 30 minutes, the number of particles dropped sharply. This reduction in the number of virus particles continued until <10% of the virus particles survived after 2 hours.

Figure 5.5 shows the number of particles counted after exposure to 70°C. Note about a 50% decline in the number count after 1 minute of exposure. Exposure up

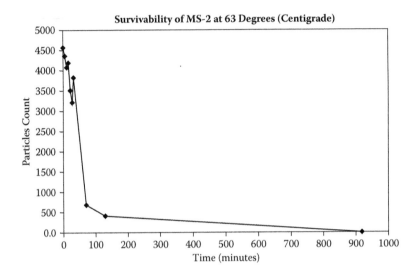

FIGURE 5.4 Number of MS2 particles as function of time at 63°C.

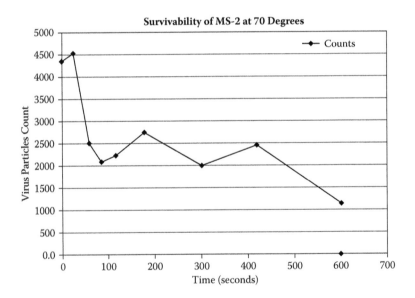

FIGURE 5.5 Survivability of MS2 at 70°C.

to 7 minutes did not change the survival rate of the virus particles. However, longer exposure (10 minutes) resulted in a sharp reduction in the survival rate.

Figure 5.6 and Figure 5.7 show the concentrations of lower molecular weight particles in four particle size bins. In both cases, an initial mix of breakdown products in all four size bins was observed. Over time, the number of particles in all four bins diminished, probably as a result of further break down to molecules of smaller weights.

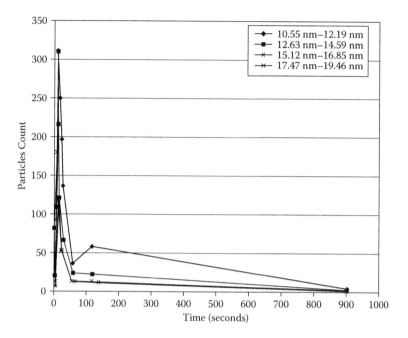

FIGURE 5.6 Concentrations of breakdown products at 63°C.

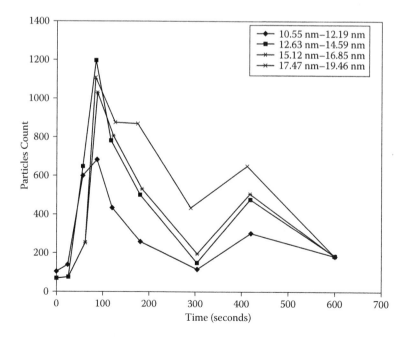

FIGURE 5.7 Concentrations of breakdown products at 70°C.

5.2.3 CONCLUSIONS

The IVDS has proven a useful tool for investigating the stability of virus particles in harsh environments. In this study, we demonstrated that MS2 can survive for a limited time (up to 30 minutes) at temperatures as high as 63°C. Longer exposures or exposures to higher temperatures lead to quick drops in the survivability of the virus particles.

5.3 DOSE RESPONSE CURVES FOR MS2 (HIGH PH)

5.3.1 PROCEDURES

The bacteriophage solution was subjected to high pH to determine the survival rate. A high pH solution was prepared by mixing 1:1 solutions of 1 M 2-aminoethanol and 20 mM ammonium acetate. The resultant solution had a pH of 11.1. For testing, 10 μL of the concentrated MS2 was mixed with 10 μL of the high pH solution, and the number concentration of virus particles was measured over time (Wick et al. 2006b).

5.3.2 RESULTS

The analysis results of the initial purified and concentrated sample of MS2 are shown in Figure 5.8. The resultant graph shows a typical MS2 peak at 22.3 nm. The counts were numerically divided by a factor of two to allow graphical comparison with the high pH solutions. Figure 5.9 provides the analysis results immediately after mixing the MS2 solution with the high pH solution. Subsequent analyses were

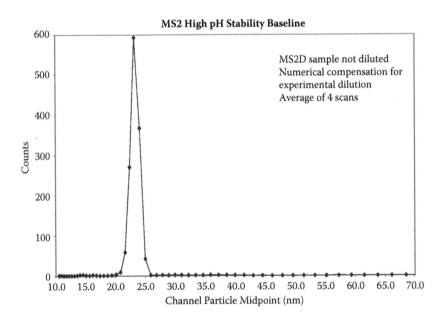

FIGURE 5.8 Baseline MS2 count.

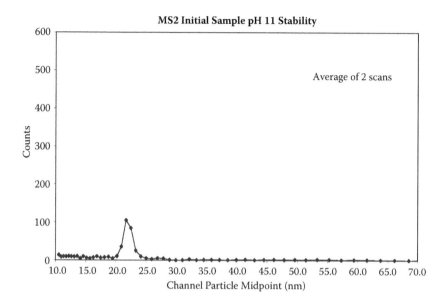

FIGURE 5.9   MS2 count immediately after addition of high pH solution.

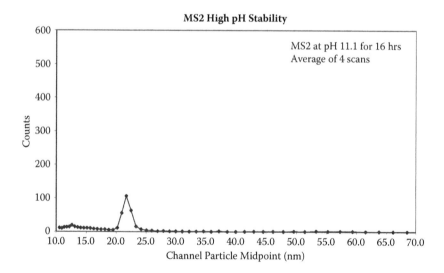

FIGURE 5.10   MS2 count after 16 hours in high pH solution.

conducted after 16 hours and after 42 hours. These results are shown in Figures 5.10 and 5.11, respectively.

By comparing Figures 5.8 and 5.9, we can see that the concentration of virus particles drops from almost 800 to about 140 immediately upon increasing the pH. However, after 16 hours in a highly alkaline solution, the concentration remains the

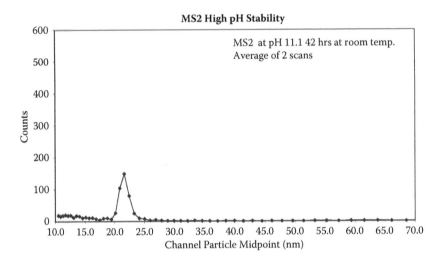

FIGURE 5.11   MS2 count after 42 hours in high pH solution.

TABLE 5.1
Average Particle Counts of MS2 (High pH)

	10.55 to 19.46 nm	20.17 to 29.96 nm	Comments
Baseline	20 ± 0.8	1355 ± 194	Initial dilution 1:1, numerical compensation to 1:2 experimental dilution
Initial sample	144 ± 2.4	286 ± 35	Diluted 50:50 from baseline
16 hours at pH 11.1	189 ± 3.7	268 ± 34	Diluted 50:50 from baseline
42 hours at pH 11.1	237 ± 4.7		Diluted 50:50 from baseline

same and even shows a sign of increasing after 42 hours. The results are tabulated in Table 5.1. As the concentration of viruses declines, we can expect that the cell material and the core of the virus will disintegrate to smaller sized protein particles. This is indeed what happens as observed in Figure 5.12 depicting the IVDS results for particles smaller than 20 nm. Figure 5.13 shows the increase in these particles after 42 hours.

5.3.3 Conclusions

Using the IVDS, we were able to show that high pH causes a breakdown of viruses. Initially, upon increase of the pH from neutral to highly basic, the concentration of virus particles diminished rapidly. Degradation then slowed over time. The degradation of viruses was accompanied by increased concentration of proteins that are breakdown materials of the cells and cores of the virus.

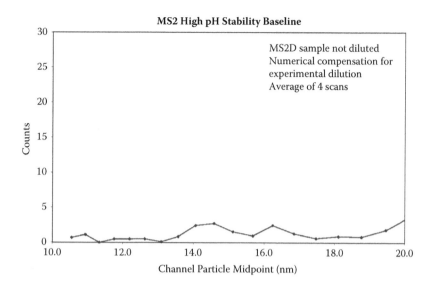

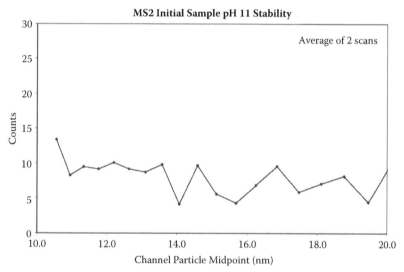

FIGURE 5.12 IVDS scan of particles smaller than 20 nm (high pH). *(contnued)*

5.4   DOSE RESPONSE CURVES FOR MS2 (LOW PH)

5.4.1   PROCEDURES

The same general methods were followed to determine temperature dose response curves for a low pH solution. The concentrated MS2 solution was added to 23 mL of 20 mM ammonium acetate solution, then concentrated again by the UF subsystem to a total of 2.5 mL of clean solution. The bacteriophage solution was then subjected to low pH to determine the survival rate and the dose response curves (Wick et al. 2006c).

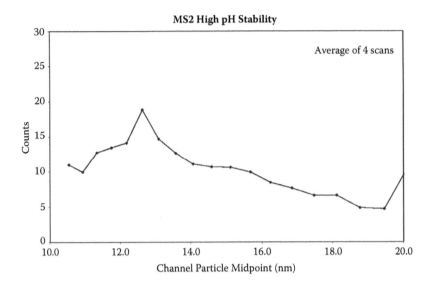

FIGURE 5.12 (continued)   IVDS scan of particles smaller than 20 nm (high pH).

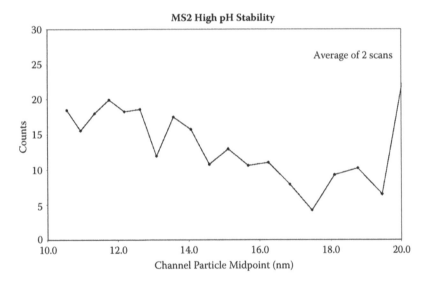

FIGURE 5.13   MS2 sample 42 hours after increasing pH.

For low pH testing, 10 μL of the MS2 stock solution was added to 90 μL 0.1 N nitric acid (HNO$_3$). The resultant solution had a pH of 1.4. Prior to scanning by the IVDS, the acidic solution was neutralized by adding a buffer solution comprised of 5.3 μL of 0.1 N ammonium hydroxide (NH$_4$OH) and 84.7 μL of 20 mM solution of ammonium acetate (pH 7.05). The low pH MS2 solution was neutralized for a few minutes after subjecting the MS2 to the low pH and again after 66 hours.

5.4.2   RESULTS

The original stock solution of MS2 was diluted 1:10 with a solution of 20 mM ammonium acetate and analyzed by the IVDS. The scan results are shown in Figure 5.14. The counts were numerically divided by a factor of 10 to graphically compare them with results after subjecting the MS2 to a low pH environment.

Figures 5.15 and 5.16 show IVDS scan results immediately after the sample was subjected to the acidic conditions and again after 66 hours, respectively. The MS2

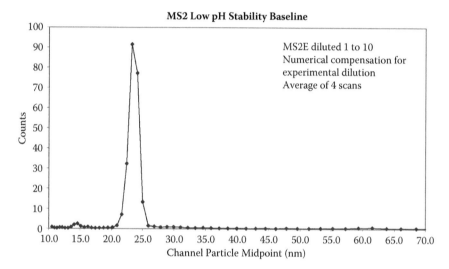

FIGURE 5.14   MS2 sample baseline count.

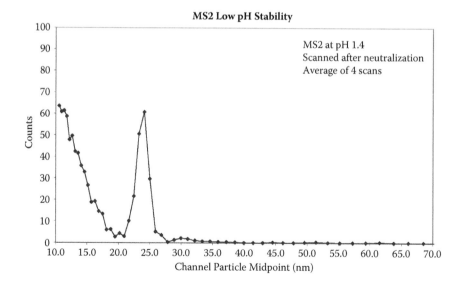

FIGURE 5.15   IVDS scan immediately after adding low pH solution.

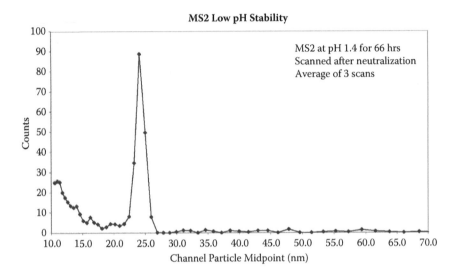

FIGURE 5.16   IVDS scan of MS2 at low pH after 66 hours.

TABLE 5.2
Average Particle Counts of MS2 (Low pH)

	IVDS Counts (Size Range of 10.55 to 19.46 nm)	IVDS Counts (Size Range of 20.17 to 29.96 nm)
Baseline	12 ± 0.6	229 ± 32
Initial sample	601 ± 21	192 ± 21
66 hours at pH 1.4	214 ± 8	202 ± 28

particle counts are roughly the same, indicating that MS2 can survive intact at low pH for many hours. The results are tabulated in Table 5.2.

Figure 5.17 shows the IVDS scan results for a particle range of 11 to 20 nm where we can expect to find degradation products of MS2. The figure shows that the counts at this size range remain fairly consistent. This is further confirmation that MS2 is stable under strong acidic conditions (pH as low as 1.4). Note that HNO_3 is considered a strong oxidative solution. The results indicate that some viruses can survive such harsh environments.

5.4.3   Conclusions

This study shows that certain viruses can survive harsh environments such as low pH and highly oxidative environments. The results of this study are of practical importance because H_2O_2 vapors are being considered as a means to decontaminate

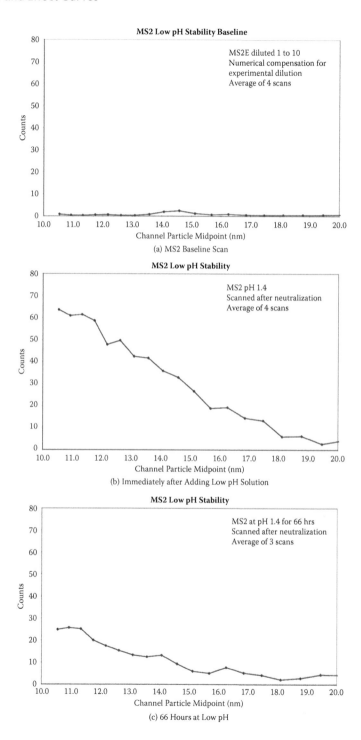

FIGURE 5.17   IVDS scan for MS2 breakdown products at low pH.

building interiors after a biological terrorist attack. Because viruses can be used as terror weapons, we recommend the examination of the fates of viruses considered potential biological weapons under such conditions. We need to emphasize we investigated only the impacts on the structural integrities of virus particles in this study. We did not test virus viabilities.

6 Isolating Single Viruses from Multivirus Mixtures

6.1 INTRODUCTION

The invention (Wick 2013a; Wick 2013b) of a means and method to isolate a single virus from a mixture is reviewed in this section. The isolation and collection of these viruses represent a new approach to the difficult isolation problem and increase our ability to achieve detailed analysis of single isolates from complex environments.

The detection and correct classification of nanometer-sized viruses that include previously unknown, emerging, or engineered strains require novel approaches and strategies. The approaches based on identification via analysis of nucleic acids using polymerase chain reaction (PCR)-based methods, gene chips, and sequencing are effective in identifying known pathogens, but development of novel approaches is needed to detect and classify unknown or engineered microbes of interest.

Current methods are limited in their effectiveness and microbes that are not sequenced may be invisible to these methods. One major issue with biomolecular approaches is that samples containing mixtures and environmental components may interfere with precise molecular biology processes. Thus, a real-world sample that contains multiple viruses, various nanometer components of cells, and several salts and metals can pose a difficult prospect for current systems.

Another point to consider is that samples containing many components require a lot of time to analyze and explain. Unknown viruses need to be isolated for further analysis, cloning, sequencing, and processing. It is difficult to have two or more unknown viruses in a sample while trying to use common methods for analysis. A new method that can detect, separate, isolate, and enhance nanometer-sized particle viruses from real-world samples is indicated. IVDS can detect and classify viruses and the invention described in this section is an improvement over other methods.

6.2 PROCEDURE

Figure 6.1 depicts the IVDS. The system includes a collection stage 101, an extraction stage 102, a purification and concentration stage 103, and a detection stage 104. These four stages are described in Chapters 1 and 2.

When a sample passes from the purification stage 103, the retentate enters the detection stage 104 at the inlet of the electrospray capillary of the electrospray assembly in the fourth position of the positioner. Entry into the electrospray capillary is achieved without passing the retentate through piping that could cause sample losses. The electrospray capillary is on the order of 25 cm in length, and the inlet of the electrospray capillary is positioned to the small front face side collection volume

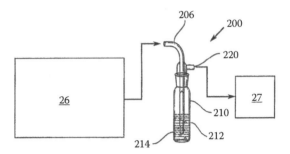

FIGURE 6.1   In-line aerosol collector.

of the ultrafiltration (UF) membrane 22. The electrospray capillary is then positioned to sample liquid from the retentate side of the filter and the sample liquid enters the electrospray assembly 24.

The liquid sample solution in the electrospray assembly passes into an orifice or jet of 50-micron diameter and droplets are ejected under the influence of an electric field. The droplets are typically between 0.1 and 0.3 microns in size, with a fairly narrow size distribution. At a droplet size of 1.3 microns, sampling rates are 50 nanoliters per minute (nl/min), allowing the electrospray assembly 24 to spray the collection volume at a rate of 20 minutes per microliter.

From the electrospray assembly, the sample passes to a charge neutralizer 25. The charge on the droplets is then rapidly recovered using an ionizing atmosphere to prevent Rayleigh disintegration. The neutralized electrospray droplets are dried in flight, leaving the target virus molecules and/or dried residue of soluble impurities. From the charge neutralizer 25, the target virus molecules and/or dried residue enter the differential mobility analyzer (DMA).

The DMA uses the electrophoretic mobility of aerosol particles to classify the particles by size, using the inverse relationship of the mobility of a particle to its size. In the DMA, particles are carried by an air stream at a set velocity through an electric field created by a charged rod. If a particle is singly and positively charged, it experiences an electrostatic attraction to the rod, which competes with the inertial force of the flow.

When the electrophoretic mobility falls within a certain range, the particles pass through a narrow exit port at the end of the charged rod. The particle size range (generally 0.01 to 1 micron) is divided into size channels. The entire range is automatically scanned in 1 to 10 minutes, generally within 3 minutes. The DMA has a possible instrumental error of only 3% for virus size determination. Additionally, a size increase may result from covering of virus particles with impurity residues. At an impurity level of 100 ppm, a typical 40 nm virus has a possible size error up to about 2%. If the impurity levels are less than 20 ppm, the error becomes smaller than 1%.

When the sizes of primary droplets from the electrospray assembly are 0.3 micron, a 1 ppm soluble impurity creates a 3 nm residue particle and a 125-ppm soluble impurity creates a 15 nm particle. Particles that are 15 nm in diameter can be separated in the DMA from viruses that are at least 22 nm in diameter. Accordingly, soluble impurities must be reduced to less than 100 ppm (0.01%) to avoid background interference with virus signals.

Detection of proteins at levels of 10^{11} to 10^{12} molecules per milliliter indicates that a sensitivity level for viruses of 10^{10} particles per milliliter or less has been achieved as described elsewhere. The description of this process is included for completeness because sample processing through the DMA is an important step. The features of the DMA allow a single virus to be separated from other viruses in the same sample. It is possible by using the DMA to validate against false positives by changing the dilution and seeing whether the particle size also changes. Additionally, the DMA can provide another layer of protection against interference from impurities up to 100-ppm level.

After a sample leaves the DMA, it enters the condensation particles counter (CPC) also known as the condensation nucleus counter (CNC) that utilizes a nucleation effect. Response times for step changes in concentration are less than 20 seconds; all components operate in a temperature range of 10 to 38°C. The system is controlled by a computer. When data collection and instrument control are handled by a single computer, the computer may vary the mode of operation in response to virus detection. This control allows selection of a single particle size or a size range in cases of overlaps.

The addition of this capability to the IVDS was intended to achieve the separation and identification of nanoparticles such as viruses. Further goals were collecting and separating other submicron- or nano-sized particles. This feature is important breakdown products discussed in Chapter 3 must be isolated for further analysis.

The IVDS can process and separate nanometer particles by size, within 4 nm resolution. In a single sample, several sized particles can be separated and characterized based on their differential mobilities. The system can be used as a basis for isolating and enriching an individual virus from mixtures of viruses or biological particulates.

As depicted in Figure 6.1, IVDS may be improved by including an aerosol collector (AC) 200 between the DMA 26 and the CPC 27; the collector is described in detail below. This improvement permits specific viruses to be collected after separation by the DMA. For example, if the DMA is set to separate the MS2 virus at 24 nm, the MS2 can then be collected in the aerosol collector as part of the IVDS. The collected sample can then be further analyzed by the IVDS before entering the CPC or subjected to further laboratory analysis by PCR or another technique.

The DMA is a component of the IVDS. It utilizes the electrophoretic mobilities of aerosol particles to classify particles by size using the inverse relationship between the mobility of a particle to its size. In the DMA, particles are carried by an air stream at a set velocity through an electric field created by a charged rod. If a particle is singly and positively charged, it experiences an electrostatic attraction to the rod, which competes with the inertial force of the flow. When the electrophoretic mobility falls within a certain range, the particles pass through a narrow exit port at the end of the charged rod. The particle size range (generally 0.01 to 1 micron) is divided into 147 size channels. The entire range is automatically scanned in 1 to 10 minutes, generally within 3 minutes.

As depicted in Figure 6.2, further improvements were made in the IVDS by incorporating an electrostatic collector (EC) 250 between the DMA and the CPC.

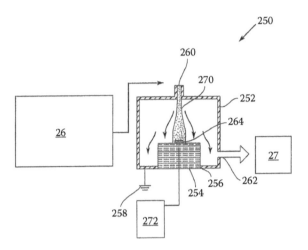

FIGURE 6.2   In-line electrostatic collector (EC).

The electrostatic collector is used to sample aerosols that have been conditioned and positively charged. The device consists primarily of a grounded 258 cylindrical sampling chamber 252 with an electrode 254 and an insulator 256 at the bottom of the chamber. The collector has an inlet 260 and an outlet 262. In the substrate 264, the aerosol flow focuses particles 270 onto substrate 264. Particles in the air flow enter the inlet into the interior of the chamber. If they are not collected on the substrate 264, they leave outlet 262 to enter the CPC.

Several experiments were performed to determine the feasibility of collection and separation of sample virus and polystyrene latex (PSL) particles. The IVDS separates particles based on its ability to monitor and control the parameters of differential mobility in the DMA while analyzing samples. Several methods including filter capture, electrostatic collection, and aerosol collection were examined to collect separated particles for further analysis.

6.3   DIFFERENTIAL MOBILITY SEPARATION

The IVDS analyzes viruses and other nanometer-sized materials by separating particles according to their mobilities in an electric field. The electrical mobility of a particle is a function of its size and the number of charges in contains. For particles in the size range of a few nanometers, the electric charge that a particle will acquire is limited to a single elemental charge forced by the passage of an aerosol stream past an ionization module. Thus, the mobility of the charged particles in an electric field will be limited to their size only. IVDS has been used to analyze viruses in various real-world matrices including air, water, soil, sand, plant and animal matter, and numerous other materials.

The DMA of the IVDS is composed of a cylinder with a central rod. A controlled DC potential differential (0 to 10,000 VDC) is applied between the cylinder and the rod. By controlling this potential, only particles with very narrow electrical mobility (size) are allowed to enter the opening slit at the bottom of the cylinder and exit the

DMA. Computer-controlled timing software ensures that particles exiting the DMA have a known electrical mobility, and they can then be counted, sized, or collected. Another possible collection area in the IVDS system is the exhaust from the condensation particles counter (CPC). Particle capture can be monitored with the IVDS computer-controlled interface. The particle display is shown in near-real time and allows control of the flow parameters in the instrument.

6.3.1 AEROSOL COLLECTION

The apparatus 200 in the center of Figure 6.1 is an All Glass Impinger AGI-30 device used to sample microbial aerosols of various types. The AGI-30 is an all-glass device that operates in the following steps. An aerosol flow enters the collector through a curved inlet tube generally simulating a human nasal passage and flows into a container 210. The container is partially filled with a liquid and the aerosol is passed into the liquid.

In this application, 80 mL of ammonium acetate buffer was used as the liquid in the container. Excess aerosol gas bubbles through the liquid and then exits the container outlet. In some applications, a vacuum is used to facilitate microbial collection. In this application, the aerosol exits the DMA component of the IVDS and is fed into the aerosol collector under pressure, usually 2 psi at a flow rate of 2 liters per minute. Collection of microbes occurs as the bubbles pass through the liquid in the container. The microbes are deposited in the liquid as the bubbles pass. In the cases of small viruses, the aerosol flow rate, bubble size, and time the bubbles spend in the liquid are important variables.

The IVDS was adapted to allow an AGI-30 aerosol collector 200 to be placed in line between the DMA and the CPC as shown in Figure 6.1. The collector was filled to 80 mL with a 20 mM ammonium acetate buffer with 50 mM isopropyl alcohol (IPA). The IVDS was set to scan at 20 to 30 nm for 30 seconds and return. The scan was repeated 60 times for a total sampling time of ~30 minutes and the MS2 bacteriophage was tested.

Figure 6.3 is a graph showing a standard solution containing MS2. The nominal particle size is 23 nm. MS2 is used as a size standard for calibration of instruments and as a standard virus particle for routine laboratory comparisons. The sample shown in the graph was diluted from a primary solution to 1:20 in 50 mM IPA in 20 mM ammonium acetate (AA) buffer. The average number of scans represents the number of separate times the IVDS instrument counted the MS2 in the sample averaged over a number of counting periods. The caption on the horizontal axis represents the particle sizes in nanometers, in this case the particle midpoints. The IPA was added to reduce the surface tension in the water.

The initial scan of the stock solution used in other experiments exhibited a well-defined MS2 peak at 24 nm. After collection in the aerosol collector 200, a sample may be tested further or evaluated by other means. After collection in the aerosol collector 200, the 80-mL sample was again analyzed with the IVDS. Figure 6.4 confirms that MS2 was collected and indicates that the collected sample contained a small amount (relative to the original sample amount) of MS2. The 80-mL sample was then ultrafiltered to concentrate the volume down to 1.5 mL. The scan of the

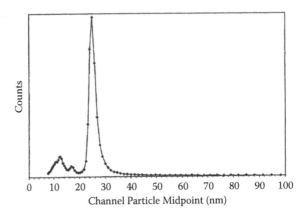

FIGURE 6.3   IVDS standard solution containing MS2.

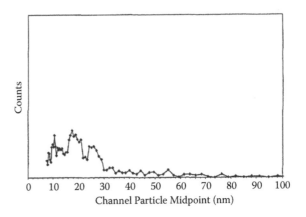

FIGURE 6.4   Verification of collection of MS2 by in-line aerosol collector.

ultrafiltered sample (Figure 6.5) shows a better defined MS2 peak at 24 nm, again confirming that MS2 was collected from the sample.

Figure 6.6 shows the particle counts for an unfiltered sample taken directly from the AGI-30 aerosol collector. The caption at the top is record keeping data indicating that the MS2 sample was collected in 50 mM IPA in AA buffer. The sample was collected while the liquid was stirred. The figure shows an average of six scans (IVDS counting periods) and represents the particle counts for the solution in Figure 6.5 that was concentrated from 80 to 1.6 mL via ultrafiltration using a 600 KDa filter.

6.3.2   ELECTROSTATIC COLLECTION

Figure 6.2 depicts the electrostatic collector used to sample an aerosol that has been conditioned and positively charged. The aerosol stream in the IVDS is positively charged by passing through an ionizing source, i.e., the DMA. The collector consists of a grounded cylindrical sampling chamber with an electrode at the bottom. A

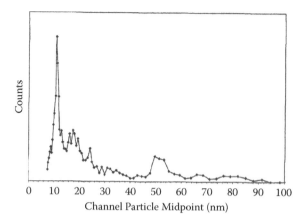

FIGURE 6.5   Further processing of sample shown in Figure 6.4 showing increased counts.

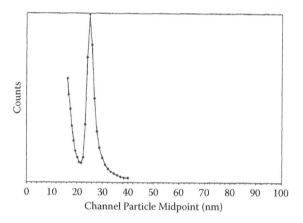

FIGURE 6.6   MS2 collection with electrostatic collector between DMA and CPC.

negative potential up to 10,000 volts can be set on the electrode. A pump is used to draw aerosol into the chamber through the inlet. In operation, a sample substrate is fixed to the electrode with adhesive tape or other means. The sampler is then run at a fixed flow rate and voltage. The electric field between the grounded chamber and the electrode focuses particles onto the substrate.

The substrate can be any electrically conductive material, for example, stainless steel foil was used for some tests. The substrate can then be removed after a time and the particles may be analyzed further. For example, the particles collected can be redispersed into a buffer solution and then analyzed by the IVDS. The arrows within the chamber indicate air flow.

A sample of MS2 (D3) was run on the IVDS with the electrostatic collector in line between the DMA and the CPC (Figure 6.2). The flow is provided by the IVDS hardware. After washing the stainless steel disk with buffer solution, the sample was analyzed by the IVDS. Figure 6.7 reveals is a small peak for MS2 after collection; confirming the collection of MS2 from the sample.

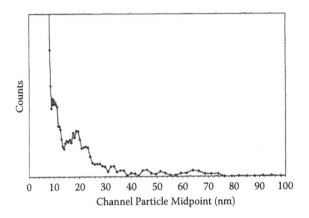

FIGURE 6.7   Confirming collection of MS2 from electrostatic collector.

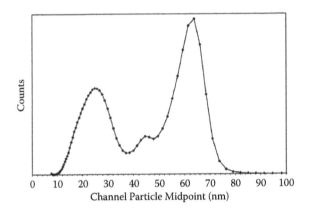

FIGURE 6.8   IVDS results of 30- and 70-nm polystyrene latex stock mixture.

6.3.3   POLYSTYRENE COLLECTION

A sample of mixed polystyrene latex (PSL) particles of 30 and 70 nm size were also collected with the electrostatic collector. The results from the stock mixtures are shown in Figure 6.8. The samples were redispersed in a buffer solution for IVDS analysis. Figure 6.9 shows the IVDS results for the 30 and 70 nm electrostatic collections.

6.4   CONCLUSIONS

The enhanced IVDS constitutes an improved system for collecting, detecting, and classifying submicron-sized particles in a mixture. The modification of the IVDS consisted of inserting an aerosol collector or an electrostatic collector device between the DMA and the CPC to allow submicronsized particles from the environment to be detected, separated, sized, and identified.

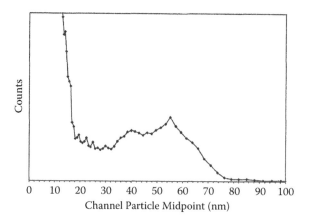

FIGURE 6.9 IVDS analysis of collected polystyrene latex sample.

The IVDS invention involves a method for detecting nanometer-sized viruses in a sample by eliminating components that interfere with the detection of the viruses. A sample suspected of containing nanometer-sized viruses is inserted into an improved IVDS modified to include an aerosol collector or electrostatic collector. Samples to be tested may consist of air, water, soil, sand, or organic matter from plants or animals. The separation resolution power is within 4 nm. The invention can also collect and further analyze nanometer-sized viruses, polystyrene latex particles, and MS2 bacteriophages.

6.5 SUMMARY

An improved IVDS system for detecting and classifying submicron-sized particles in sample taken from the environment was devised by adding an electrostatic collector in line between the DMA and the CPC. The collector is connected to and receives the outlet air flow from the DMA and produces a grounded sampling.

The examples used in this chapter are only representations of the capabilities of the IVDS. If larger concentrations of isolated viruses or particles are required, other techniques such as longer collection periods, higher initial concentrations, or increased concentration via ultrafiltration may be employed. Combinations of methods can be used to increase the starting concentrations of target particles to speed the collection of the isolated particles, for example, adjusting IVDS settings, using different capillaries, and other such means.

The IVDS presents noteworthy implications for the science of virology. It is now possible via mechanical means to isolate a single virus from a complex containing many viruses. The isolates can be used for a variety of applications. Unknown particles can be analyzed further to determine their identifications and other features using such techniques as mass spectrometry proteomics, genome sequence, and related methods.

7 Analyzing Various Viruses

7.1 INTRODUCTION

Viruses are considered to be among the smallest living particles known to humans. Figure 3.1 in Chapter 3 illustrates the comparative sizes of different types of particles. Because of their small size, viruses are extremely difficult to detect and characterize. Since they can cause many diseases in humans and can also affect crops and domestic animals, they are of great concern to public health authorities because they can exert major economic impacts by causing plant and animal diseases.

Detection and identification of viruses are presently complex and expensive biochemical processes that require great expertise. Even new methods that rely on the newest technology are essentially biochemical procedures. The detection process is particularly complex and lengthy when an unknown virus "hits the street." This was illustrated clearly by the length of time it took to discover and identify the HIV virus.

As mentioned in earlier chapters, IVDS was developed and invented at the Edgewood Chemical and Biological Center (ECBC) and it allows viruses to be detected and identified by physical rather than biochemical means. The IVDS relies on the varying sizes of different viruses. The system isolates the virus particles from extraneous material in which they are collected, separates them according to their sizes using a differential mobility analyzer (DMA), and determines their concentrations via a condensation particles counter (CPC). The system allows quick screening of many samples at low cost. Using the IVDS, we developed detailed procedures to isolate and characterize many viruses. In this chapter, we describe the details and procedures by which we isolated and characterized the viruses (Wick et al. 2005d).

7.2 IVDS

The IVDS is described in detail in Chapter 2 and its operation is covered in Chapter 1. We include a summary here of the basic principles of operation of the IVDS and its various components. For more detail, readers are also referred to other chapters for information about its operation, filtration steps, and other factors. The IVDS is a modular system built around four principal components:

- Purifier/concentrator
- Electrospray injection
- Size analyzer composed of differential mobility analyzer (DMA)
- Condensation particles counter (CPC) used to determine concentration

The IVDS is designed as a modular system to allow upgrades of different components as they are developed and improved, for example, the additions allowing isolation of single viruses discussed in Chapter 6.

7.2.1 PURIFIER/CONCENTRATOR

The ultrafiltration (UF) system is used as its name implies to remove extraneous material from a sample and concentrate the viruses to a point where they can be detected. The UF system consists of a fiber-based tangential flow filtration system where the more coarse contaminations are removed and the sample volume is reduced to less than 1 mL. The initial sample volume can range from few milliliters to several liters.

The most important component of this stage is a bundle of hollow fibers whose walls are made of a permeable membrane with a wide range of pore sizes. The sample is pumped through the permeable fiber and the filtrate is forced through the fiber walls by the pressure differentials. The sweeping action of the sample stream prevents clogging of the fibers. Fibers with pore sizes from 4 μm to 100 KDa are available.

7.2.2 ELECTROSPRAY INJECTION

The sample solution is stored in a cone-shaped vial enclosed in a cylindrical pressure chamber. The chamber accommodates the inlet capillary and a platinum high-voltage wire, both of which are immersed in the solution. Maintaining a differential pressure causes the solution to be pushed through the capillary. The fluid containing the particles exits the capillary and is sprayed through a strong electric field that causes it to form a cone and break into small charged droplets. To prevent corona discharge, the cone is surrounded by CO_2. As the droplets evaporate and dry, they form a plume of particles consisting of the virus-like particles (or other large molecules) contained in the original sample plus solute salt particles. If the concentration of the virus-like particles is not too high, each particle will contain a single virus.

7.2.3 DIFFERENTIAL MOBILITY ANALYZER (DMA)

The DMA separates particles according to their mobilities in an electric field. The electrical mobility of a particle is a function of its size and the number of charges it contains. Particles in the size range of few nanometers are limited to acquiring single elemental charges. Thus the mobility of the charged particles in an electric field will be based on their size only.

A plume of polydispersed particles enters the DMA consisting of a cylinder with a central rod. A controlled DC potential differential (0 to 10,000 VDC) is applied between the cylinder and the rod. By controlling this potential, only particles with very narrow electrical mobility (size) can pass through the opening slit at the bottom of the cylinder and enter the particles counter. Figure 7.1 illustrates the resolution capabilities of the IVDS. A mixture of rice yellow mottle virus and MS2 was

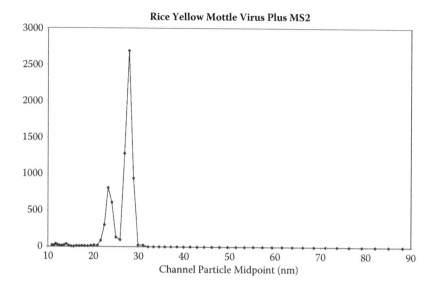

FIGURE 7.1 The 4 nm resolution of IVDS-DMA using rice yellow mottle virus and MS2.

prepared in the laboratory and analyzed. The peaks of the two viruses (4 nm apart) are clearly separated.

7.2.4 CONDENSATION PARTICLES COUNTER

The particle counter is based on the need of a vapor to reach a high degree of supersaturation in order to condense and form droplets in the absence of any particles. Water vapors, for example, need to reach a supersaturation of 800% before spontaneous (also known as homogeneous) nucleation occurs. Small particles serve as nucleation (condensation) centers and condensation occurs at a lower supersaturation level which is a function of particle size. It can be shown that all the nucleating particles grow to droplets of identical size (depending on the availability of vapors). The number of particles then is identical to the number of droplets formed. This number can be deduced by measuring the opacity (or light transmission) in the volume containing the particles.

7.3 SAMPLE ANALYSIS PROCEDURES

7.3.1 GENERAL PRINCIPLES

The process for analysis of a sample for viruses is described in Figure 7.2. The analysis process depends on the quality of the sample. If a sample is clean and concentrated enough, the process is straightforward. However, if a sample contains impurities and/or is not concentrated enough, it must undergo purification and concentration processes to make it suitable for analysis.

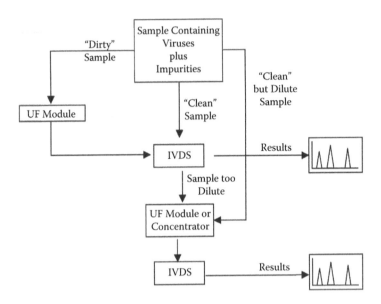

FIGURE 7.2   Logic tree for detecting viruses.

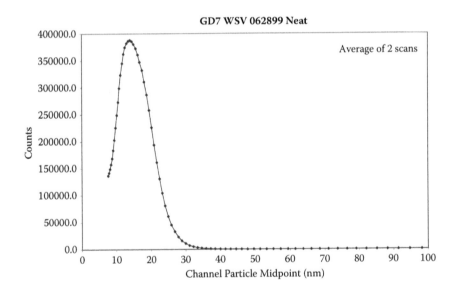

FIGURE 7.3   IVDS analysis of GD7 picornavirus.

Figure 7.3 shows the IVDS's output when injected with a sample of GD7 picornavirus received from Charles River Laboratories in Wilmington, Massachusetts (WSV 062899). The large symmetrical peak is characteristic of a large concentration of salt. After diluting the sample 1:100 with 20 mM ammonium acetate, reconcentrating it via the ultrafiltration subsystem, and rerunning with multiple scan accumulation, we

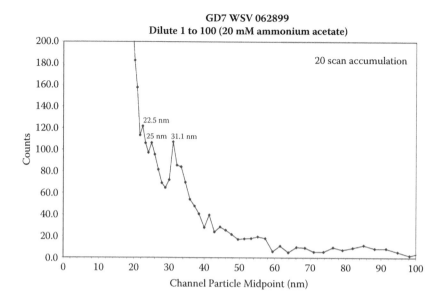

FIGURE 7.4 Ultrafiltration sub-system and rerunning with multiple scan accumulation. Peak can be seen clearly at 31.1 nm.

clearly see a peak at 31.1 nm (Figure 7.4). Full-scale expansion also shows a large peak at 15.1 nm (Figure 7.5) which may be protein or cellular debris.

A virus has several components of different sizes that can aid in its identification. Figure 7.6 shows a scan of an alphavirus. The intact virus has a peak at 70 nm. Doublets and triplets are seen at about 95 and 107 nm, respectively. In addition we see large signals at 49 and 36 nm that are likely the stripped outer protein and core material, respectively. Another example can be seen in Figure 7.7, which shows the intact MS2 virus (27 nm) and its building blocks—the RNA core (18 nm) and the protein coat (13 nm).

7.4 ANALYSES OF AND RESULTS FOR SPECIFIC VIRUSES

7.4.1 GD7 VIRUS

The GD7, also known as picornavirus-icosahedral, is a murine encephalomylitis virus. It was obtained from Charles River Laboratories in a solution containing salts and growth media. The approximate virus size was 22 to 30 nm. Figure 7.3 is a scan of the original solution; the large peak from the salt masks any virus peaks. The virus solution was diluted at a ratio 1:100 with 20 mM ammonium acetate. A multiple accumulation scan of the diluted solution (Figure 7.4) shows a clear peak at 31.1 nm and companion peaks at 25 and 22.5 nm that are likely other viruses or protein breakdown products. Full expansion shows large peaks at 15.1 and 9 nm (Figure 7.5). These are likely characteristic protein breakdown products or cellular debris.

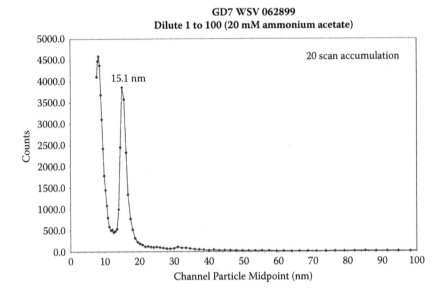

FIGURE 7.5 Full-scale expansion of sample shown in Figure 7.4 also shows large peak at 15.1 nm.

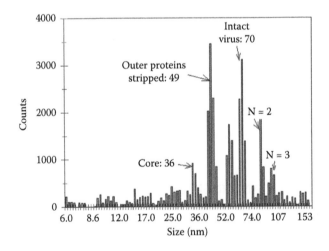

FIGURE 7.6 IVDS analysis of alphavirus. Intact virus reveals peak at 70 nm.

7.4.2 KILHAM RAT VIRUS (KRV)

The virus (Lot 091991) in salt solution was obtained from Charles River Laboratories. A run of the neat solution (Figure 7.8) shows a large symmetrical peak that is likely the salt peak. After diluting the solution 1:100 with 20 mM ammonium acetate, ultrafiltration with a 100KDa filter and rerunning with multiple scan accumulation,

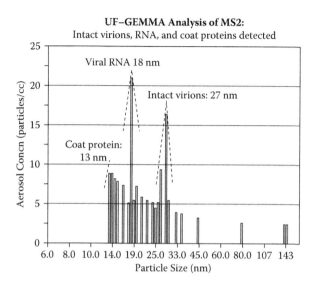

FIGURE 7.7   MS2 virus and its building blocks: RNA core (18 nm) and protein coat (13 nm).

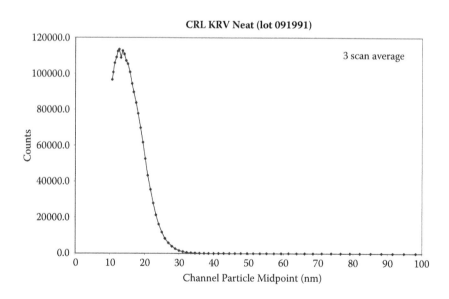

FIGURE 7.8   KRV neat solution shows large symmetrical peak that is likely a salt peak.

we achieved the results shown in Figure 7.9. The main peak of the KRV appears at 22.5 nm with companion peaks (at much lower concentrations) at 34.6, 49.6, and 71 nm, respectively. These are likely particles that contain two or more viruses each or contamination from other virus particles. A large peak at 13.6 nm is likely a protein breakdown product or cellular material not removed by filtration.

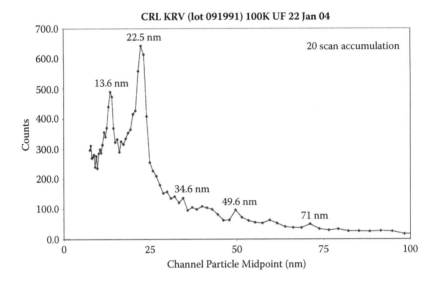

FIGURE 7.9   Main peak of KRV virus appears at 22.5 nm with companion peaks (at much lower concentrations) at 34.6, 49.6, and 71 nm. These are likely particles that contain two or more viruses each or contamination from other virus particles.

7.4.3   MAD-K87 ADENOVIRUS

The MAD-K87 adenovirus (WSV 072098) in a salt solution was received from Charles River Laboratories. A run of the neat solution shows a main virus peak at 82 nm and a companion peak at 57.3 nm, probably from a protein breakdown product (Figure 7.10). After 1:100 dilution with 20 mM ammonium acetate to remove the salt peak, the analysis shows a peak at 25.9 nm (Figure 7.11). Full-scale visualization (Figure 7.12) shows unknown peaks at 15.1 and 10.9 nm that are probably protein breakdown products.

7.4.4   MOUSE HEPATITIS VIRUS (MHV) OR CORONAVIRIDAE

MHV, also known as Coronaviridae is an enveloped virus. Lot 120998 of the virus in salt solution was received from Charles River Laboratories. Figure 7.13 shows the enveloped virus at 73.7 nm and the uncoated virus at 51.4 nm. The virus is unstable and breaks apart when diluted 1:20 with 20 mM ammonium acetate or 50 mM potassium phosphate (Figure 7.14 and Figure 7.15, respectively).

7.4.5   MVM PARVOVIRUS

A non-enveloped MVM parvovirus in salt solution was obtained from Charles River Laboratories. A scan of the neat solution (Figure 7.16) could not resolve the virus peak due to the large salt peak. Diluting the solution 1:100 in 20 mM ammonium acetate revealed a virus peak at 26.9 nm (Figure 7.17). Figure 7.18 shows protein breakdown products at 15.1 and 10.9 nm.

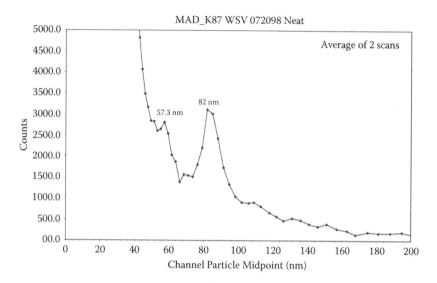

FIGURE 7.10 MAD-K847 neat solution shows main virus peak at 82 nm and unknown companion peak at 57.3 nm.

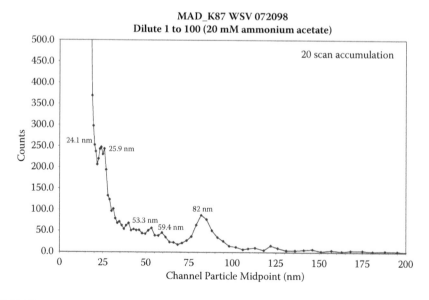

FIGURE 7.11 A 1:100 dilution of sample shown in Figure 7.10 with 20 mM ammonium acetate to remove salt peak.

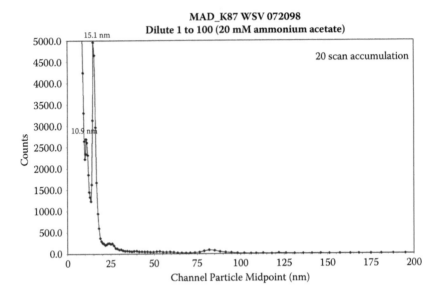

FIGURE 7.12 Further analysis of Figure 7.11 sample at full-scale visualization shows additional peaks of unknowns at 15.1 and 10.9 nm.

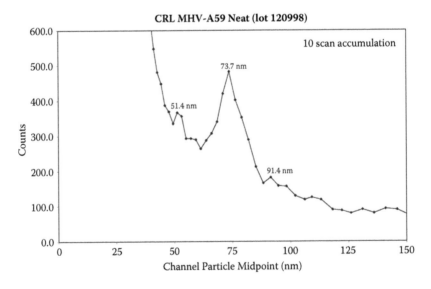

FIGURE 7.13 MHV neat shows enveloped virus at 73.7 nm and possible uncoated core at 51.4 nm.

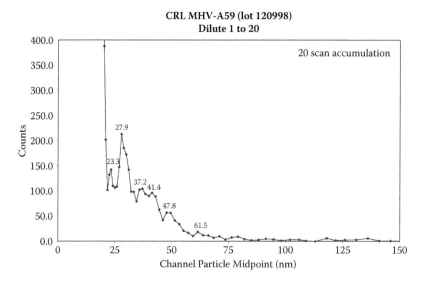

FIGURE 7.14 MHV is unstable and breaks apart when diluted 1:20 with 20 mM ammonium acetate.

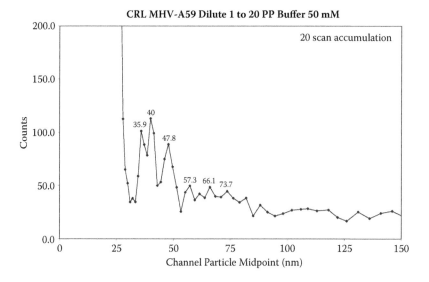

FIGURE 7.15 MHV is unstable and breaks apart when diluted with 50 mM potassium phosphate.

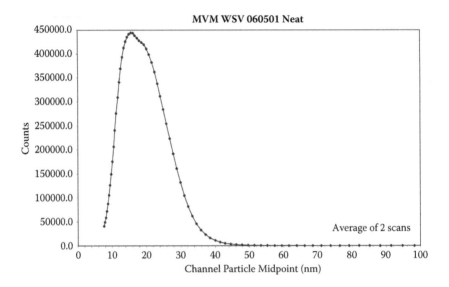

FIGURE 7.16   MVM parvovirus neat solution contains large salt peak.

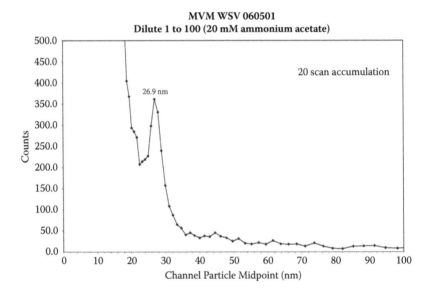

FIGURE 7.17   Diluting Figure 7.18 solution 1:100 in 20 mM ammonium acetate reveals virus peak at 26.9 nm.

7.4.6   REO-3 REOVIRUS

The Reo-3 reovirus in salt solution was received from Charles River Laboratories. A scan of the neat solution (Figure 7.19) shows the virus peak at 79.1 nm and a possible viral core at 53.3 nm. A 1:100 dilution with 20 mM of ammonium acetate shows a possible breakdown peak at 24.1 nm (Figure 7.20) and 15.1 nm (Figure 7.21).

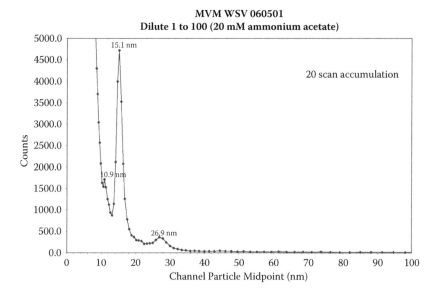

FIGURE 7.18   Two unknown peaks at 15.1 and 10.9 nm were discovered in solution from Figure 7.19.

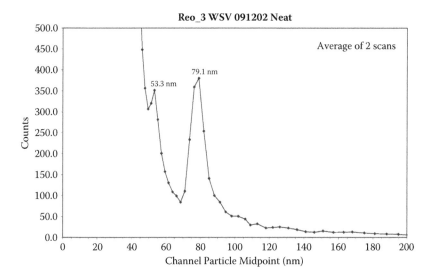

FIGURE 7.19   Reo-3 reovirus analysis of neat solution shows peak at 79.1 nm and unknown at 53.3 nm.

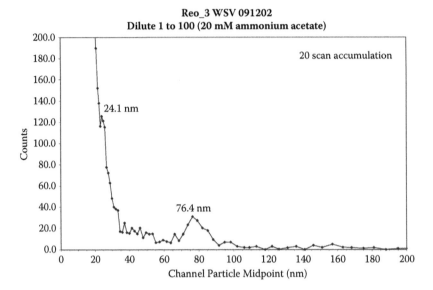

FIGURE 7.20   A 1:100 dilution of Figure 7.19 sample with 20 mM ammonium acetate shows unknown peak at 24.1 nm.

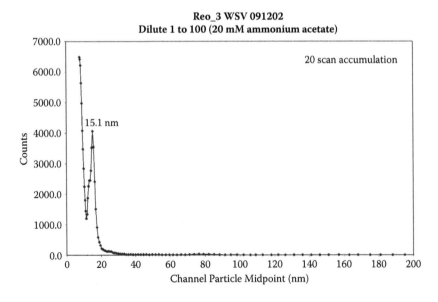

FIGURE 7.21   Solution in Figure 7.20 also shows unknown peak at 15.1 nm.

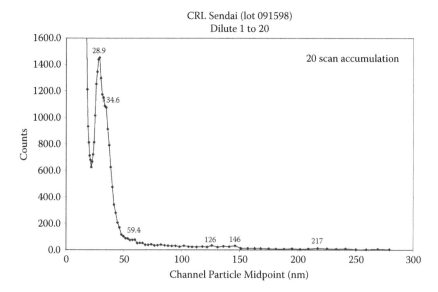

FIGURE 7.22   Sendai rodent virus diluted 1:20 with 20 mM ammonium acetate shows major peak at 28.9 nm.

7.4.7   Sendai Rodent Virus

A scan of the Sendai rodent virus diluted 1:20 with 20 mM ammonium acetate is shown in Figure 7.22. A major peak is evident at 28.9 nm.

7.5   ANALYSIS OF COMPLEX MEDIA

7.5.1   Detection of Viruses in Drinking Water

Figure 7.23 illustrates the ability of the IVDS to detect viruses in complex media. In this case, 1 mL of mixed media MS2 was mixed in 500 mL of drinking water. The sample was processed through the UF subsystem and volume was reduced to 1.2 mL and analyzed. The MS2 peak is clearly visible.

7.5.2   Recovery of MS2 from "Spiked" Blood Plasma

Figure 7.24 shows the ability of the IVDS to detect viruses in complex media such as blood plasma. Plasma from cynomolgus monkeys was spiked with MS2. The spiked solution was diluted 1:100 with 20 mM ammonium acetate buffer solution, ultrafiltered through a 100 KDa fiber filter and scanned. The MS2 peak at 22.3 nm is clearly visible. A distinct peak around 15 nm is likely a protein peak.

7.5.3   Analysis of Marine Water

Seawater is known to be rich in biologically active materials. Figure 7.25 is the IVDS scan of a sample of marine water. Four liters of water were collected from Rehoboth

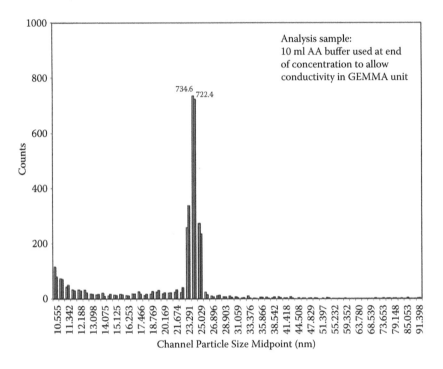

FIGURE 7.23 Ability of IVDS to detect MS2 from a "complex" medium (500 mL of drinking water).

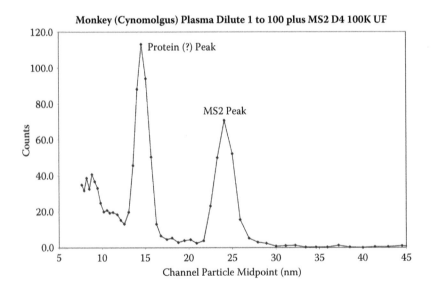

FIGURE 7.24 MS2 recovered from cynomolgus monkey plasma.

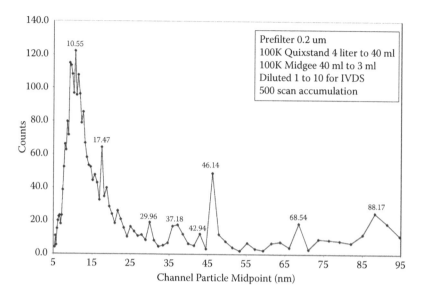

FIGURE 7.25   IVDS analysis of sample of marine water.

Beach, Delaware. The 4 liters were concentrated to 40 mL and then to 3 mL by the UF subsystem. To reduce salt peaks, the solution was diluted 1:10 with a 20 mm buffer solution of ammonium acetate and scanned by the IVDS. Figure 7.25 represents a compilation of 500 scans. Several unidentified peaks between 29 and 89 nm are clearly discernible; the peaks between 10 and 20 nm are likely proteins or cell fragments. The figure also illustrates the complexities of marine aerosols and further emphasizes the need to better examine the biological background of the marine environment as part of a strategy to detect possible bioterrorism attacks from the sea.

7.5.4   Analysis of Saliva

Figure 7.26 shows an IVDS analysis of human saliva. The saliva was mixed with distilled water, filtered through a 100 KDa fiber in the UF module, and finally mixed with 10 mL of 20 mM ammonium acetate and scanned. No viruses can be observed. However, several peaks between 5 and 25 nm are observed. These likely represent proteins or other products.

7.5.5   Analysis of Environmental Air Samples

Air samples were collected from the Edgewood area of the Aberdeen Proving Ground in Maryland using the Spincon 3. The collecting aqueous media were prefiltered (0.22 μm filter) and then passed through 100 KDa fiber in the UF module of the IVDS, washed with 30 mL of 20 mM ammonium acetate, and analyzed. Figure 7.27 shows the scan results. Several peaks are shown clearly between 5 and 70 nm.

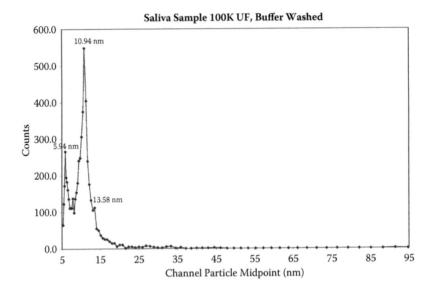

FIGURE 7.26   IVDS analysis of human saliva.

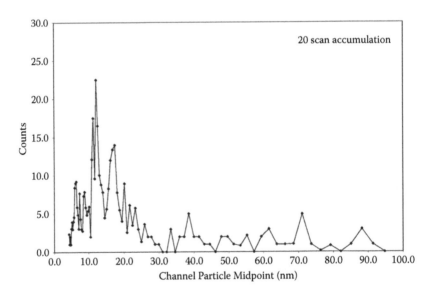

FIGURE 7.27   IVDS results of environmental air sample. Several peaks between 5 and 70 nm are clearly observed.

7.6 WATER SAMPLE ANALYSIS

7.6.1 INTRODUCTION

Detecting viruses was an initial interest following the invention of IVDS. Water is probably the most important natural resource for sustaining life forms of all types. The various materials found in water everywhere made it a likely medium for testing the performance of the IVDS. Furthermore, the need to monitor drinking water to protect it from virus hazards is becoming critical.

We used the IVDS for rapid monitoring and classification of viruses in water samples. Initial protocols for determining the presence of viruses in the various waters were considered. However, the IVDS is a generic virus detector that requires only common reagents. The system allows the detection and testing of large volumes and multiple samples of water in a short time.

7.6.2 WATER SAMPLE ANALYSIS

The types of water used as starting feed streams in the virus recovery experiments were reverse osmosis (RO), RO with chlorine (RO-Cl), and standard tap water. In addition, several native water samples were collected and analyzed with the IVDS. Examination of several feed streams determined whether contamination or constituents in the standardized water types would interfere with virus recovery and analysis.

The MS2 bacteriophage was added to the RO, RO-Cl, and tap waters to determine detection and recovery feasibility. MS2 is stable for extended periods (months), is not considered pathogenic, and has been used in many viral simulant studies.

The IVDS uses electrospray injection, a differential mobility analyzer (DMA), and a condensation particles counter (CPC) to analyze viruses. The IVDS output can be tailored to a variety of PC-based reporting software suites. A more comprehensive look at the IVDS can be found in Chapters 1 and 2.

7.6.3 CHLORINATED WATER SAMPLES

A sample of MS2 bacteriophage was added to a standard chlorinated water sample supplied by ECBC. The purpose was to determine the ability to recover the MS2 and analyze the reduction (fate) of virus concentrations in solution. The standard chlorinated water sample contained 1.5 to 2 ppm of free Cl in RO water, as measured on the morning it was collected. The sample of MS2 was analyzed before it was added to the chlorinated water.

Figure 7.28 shows the analysis of the MS2 with its resultant ~250 particle count in the 21.67 to 25.9 nm range in the ROI (region of interest). The IVDS scan of the initial chlorinated water sample (Figure 7.29) shows no virus or other contamination in the sample. After adding 0.5 mL of the MS2 sample to 1 liter of chlorinated water on the same morning the water was collected, an aliquot was taken and analyzed with the IVDS (Figure 7.30). Again, no contamination was present, and the concentration of MS2 in the sample was too low to be detected.

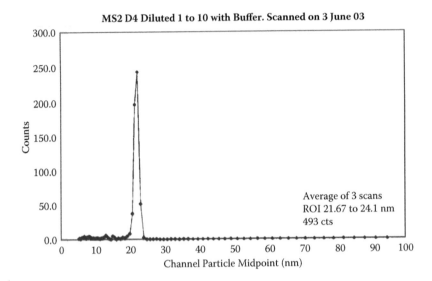

FIGURE 7.28   IVDS analysis of chlorinated water containing MS2 with resultant particle count of about 250.

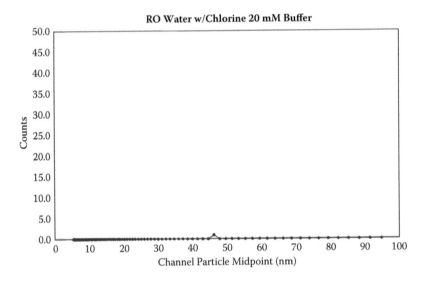

FIGURE 7.29   IVDS shows no other viruses or contamination.

The entire liter of spiked water was ultrafiltered for ~60 minutes to reduce and concentrate the volume from 1 liter to 1.2 mL. The remaining volume was analyzed by the IVDS. The reduction and concentration of the sample from 1 liter to 1.2 mL allowed the IVDS to detect the MS2 in the remaining volume. As shown in Figure 7.31, the MS2 peak shows 3 counts in the 21.67 to 25.9 nm range in the ROI. The large peaks at 16.3 and 17.5 nm are indicative of other detections and possible protein breakdown products from MS2 after disruption by chemical treatment.

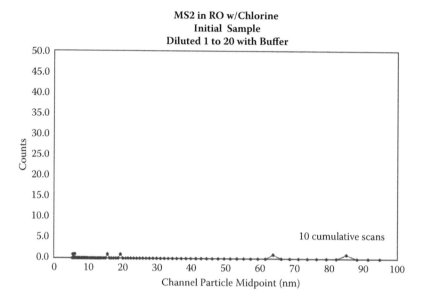

FIGURE 7.30 IVDS analysis showing addition of 0.5 mL of MS2 sample to 1 liter of chlorinated water.

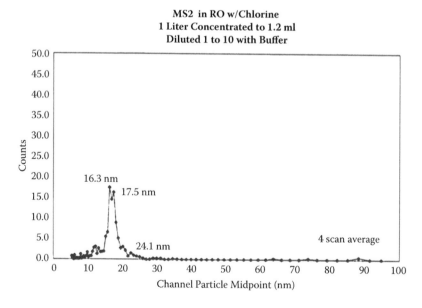

FIGURE 7.31 By concentrating sample in Figure 7.32 from 1 liter, 1.2 mL showed detection of MS2.

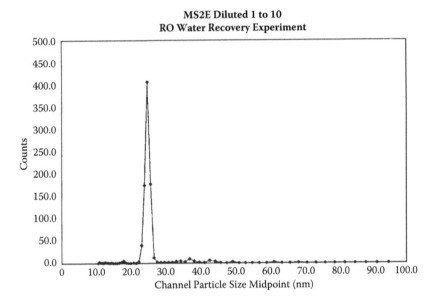

FIGURE 7.32 When MS2 was added to standard reverse osmosis water, IVDS analysis showed particle count around 400.

7.6.4 STANDARD REVERSE OSMOSIS SAMPLES

A sample of MS2 bacteriophage was added to a standard RO water sample to determine the ability to recover the MS2 from solution. The sample was analyzed before addition to the RO water. Figure 7.32 shows the analysis of the MS2 with a particle count around 400 in the 21.67 to 25.9 nm range in the ROI. The IVDS scan of the initial RO water sample in Figure 7.33 shows no viral or other contamination in the sample.

    After adding 1.0 mL of the MS2 sample to 2 liters of RO water, an aliquot was taken and analyzed by the IVDS (Figure 7.34). Again, no contamination was present, and the MS2 was in too low a concentration to be detected. The entire volume of spiked water was ultrafiltered to reduce and concentrate the volume from 2 liters to 0.7 mL. The remaining volume was analyzed by the IVDS. The reduction and concentration of 2 liters to 0.7 mL allowed the IVDS to detect the MS2 in the remaining volume. As shown in Figure 7.35, the MS2 peak shows a count of 200 in the ROI.

7.6.5 NATURAL WATER SAMPLES

Viruses occur in the natural water environment in quantities that can vary widely with conditions, seasons, and environmental changes. Several samples were collected and analyzed by the IVDS. In general, the samples were concentrated with the tangential flow (ultrafiltration) filter system before IVDS analysis to concentrate the samples to make the analysis more efficient and reduce the need to generate long accumulation scans.

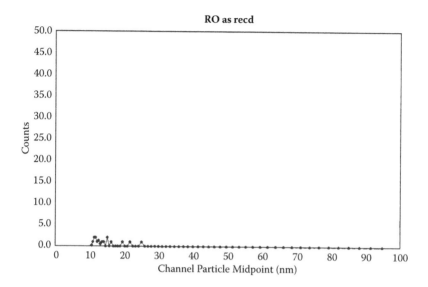

FIGURE 7.33 Standard initial sample of reverse osmosis water.

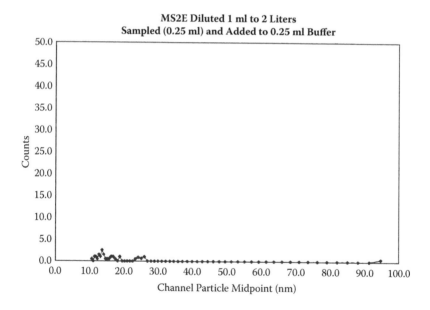

FIGURE 7.34 After adding 1.0 mL of MS2 sample to 2 liters of reverse osmosis water, IVDS showed no contamination. Concentration of MS2 was below detection level.

Figure 7.36 and Figure 7.37 show the IVDS scans from the natural water samples after filtration and concentration. The small peaks may indicate the presence of viruses in the samples. A larger study with periodic sampling at repeat sites would establish the variations in virus loading in the samples in comparison to differing environmental variables. Table 7.1 shows the volume reductions and filters used.

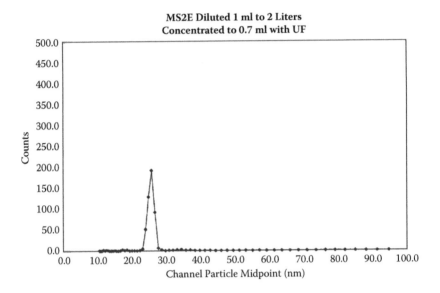

FIGURE 7.35 Sample in Figure 7.34 was ultrafiltered to reduce and concentrate volume from 2 liters to 0.7 mL. MS2 was detected easily.

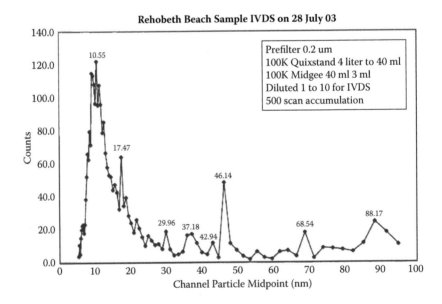

FIGURE 7.36 IVDS analysis of seawater at Rehobeth Beach, Delaware.

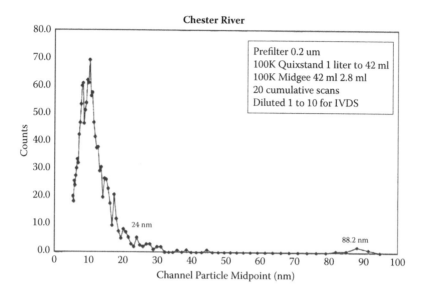

FIGURE 7.37   IVDS analysis of water from Chester River.

TABLE 7.1
Volume Reductions and Filters Used for Pretreatment
of Natural Water Samples

Filter Size	Filter Surface Area (m²)	Volume Change	Model No.
Rehobeth Beach			
0.2 μm	0.011	4 L – 4 L	3M
100KDa MWCO	0.14	4 L – 42 mL	4X2M
100KDa MWCO	0.0026	40 mL – 3 mL	MM
Chester River			
0.2 μm	0.011	1 L – 1 L	3M
100KDa MWCO	0.014	1 L – 42 mL	3M
100KDa MWCO	0.0026	42 mL – 2.8 mL	MM
Choptank River			
0.2 μm	0.011	1 L – 1 L	3M
100KDa MWCO	0.014	1 L – 40 mL	3M
100KDa MWCO	0.0026	40 mL – 3 mL	MM
Sandy Point			
0.2 μm	0.011	2 L – 2 L	3M
100KDa MWCO	0.14	2 L – 44 mL	4X2M
100KDa MWCO	0.0026	42 mL – 3.5 mL	MM

Note: MWCO = molecular weight cutoff.

7.6.6   Discussion

IVDS recovered viruses from most types of water used in this study. The reverse osmosis (RO) and tap water recovery and concentration of viruses were successful. The tap seemed to reduce some of the MS2 in solution. Further tests with varying concentrations of the constituents of the tap water may determine their effects on MS2 recovery. The RO-Cl water seemed to break down the MS2 and leave only virus fragments that were analyzed after concentration. The Cl-containing water needed subsequent analysis to determine the concentrations of Cl that are effective in removing the MS2 from solution. Various concentrations should be tested to determine the effects of Cl on MS2 and possibly other virus types.

The IVDS system was able to analyze the large volumes of water in a fairly short time. It reduced the sample volume from 2 liters to <2 mL of solution in less than 2 hours. In fact, the actual sampling time for virus analysis is less than 5 minutes. The remainder of the time was for reducing and concentrating the large volumes for analysis.

Although the IVDS only shows virus size and does not yield a definitive identification, virus types can be classed by size, thus making preliminary identification of virus type possible. Several virus types could be analyzed in various water types or in a single sample to simulate real-world conditions. After filtration, virus analysis takes only a few minutes. This confirms that the IVDS is a very fast screening tool for initial virus identification and detection. It determined that virus materials were probably present in the natural water samples tested. Further tests after identification could provide a temporal view of natural virus loading.

The ability to analyze several viruses in a single sample with simple processing makes the IVDS useful for monitoring and screening virus-sized particles in all types of water samples. In addition, if a sample (e.g., seawater) contains numerous viruses, the IVDS can determine a baseline for a given area after a few samples. The counts can be followed over time to measure natural variations and detect changes requiring alerts to monitoring personnel. The IVDS is also useful for determining fate curves for viruses in water under various conditions of temperature, pH, and other factors.

This technique is indicated for detecting viruses and virus-like particles in a variety of samples and particularly in water. It can detect several of these particles in a sample and indicate their relative concentrations. This is particularly useful for working with unknown or unexpected nanometer particles. The IVDS performed well with water from a wide range of sources—drinking water, environmental fresh water, seawater, and other types.

7.7   RECOVERY OF VIRUSES FROM SURFACES

7.7.1   Introduction

Viruses are known to survive in environmental settings over various time spans. Certain viruses survive for 24 hours; some do not survive that long, depending on the substrate with which they are in contact. This section reviews the recovery of virus samples from various surfaces (Wick et al. 2010e). The viruses were placed on surfaces and allowed to sit overnight. The sample areas were swabbed with a buffer solution and tested after 24 hours.

7.7.2 IVDS COMPONENTS

The virus recovery samples were analyzed using the IVDS described in detail in Chapters 1 and 2. A short review is included here for complexness. The detection stage of the IVDS consists of an electrospray unit to inject samples into the detector, a differential mobility analyzer (DMA), and a condensate particles counter (CPC).

7.7.3 TEST PROCEDURE

Two viruses, MS2, a bacteriophage and tomato bushy stunt virus (TBSV) were pipetted onto several surfaces overnight. The samples dried for 24 hours and were visible only from the markings placed on the surfaces before the samples were pipetted.

The sample volume for each application was 50 μL of each baseline sample (1) MS2, live TBSV, diluted 1:100 in 20 mM ammonium acetate and (2) MS2 plus TBSV. The samples were analyzed by the IVDS as shown in Figure 7.38. The peak for the MS2 bacteriophage is shown at 24.1 nm, and the peak for TBSV is shown at 32.2 nm. At the end of the 24-hour period, 50 μL of a 20 mM ammonium acetate solution was pipetted onto the virus spot and worked with a sterile cotton swab. Each swab was placed into a closed tube with 500 μL of the 20 mM ammonium acetate solution, capped, vortexed for 10 to 20 seconds, then analyzed by the IVDS.

The samples were applied to glass (microscope slide), stainless steel (balance plate), plastic (smooth computer top), ceramic (stir plate surface at ambient

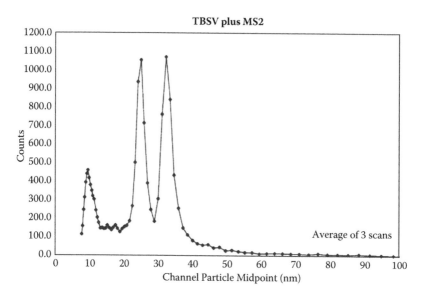

FIGURE 7.38 Baseline measurements for MS2 and TBSV.

temperature), and lacquered wood (meter stick). All sample surfaces were cleaned before application.

7.7.4 RESULTS

MS2 was recovered from the stainless steel and detected at reduced levels (Figure 7.39). Plastic results are shown in Figure 7.40, ceramic results in Figure 7.41,

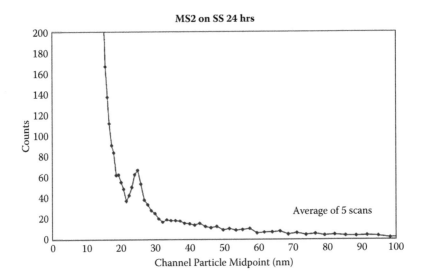

FIGURE 7.39 MS2 recovery from stainless steel.

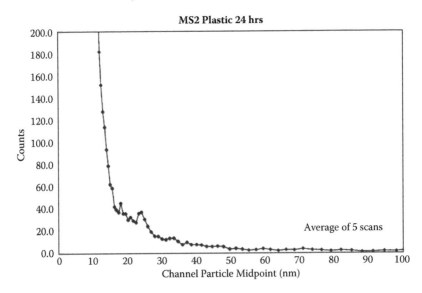

FIGURE 7.40 MS2 recovery from plastic.

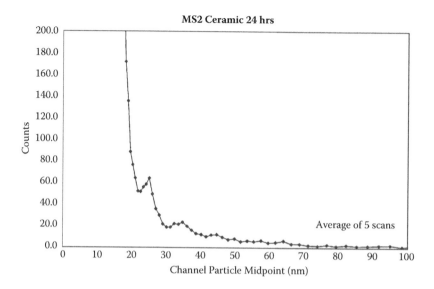

FIGURE 7.41   MS2 recovery from ceramic.

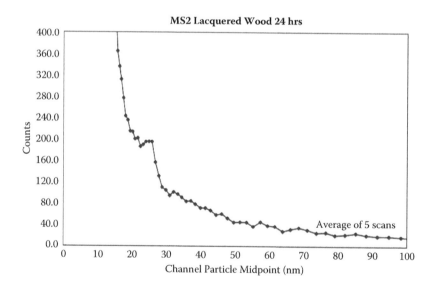

FIGURE 7.42   MS2 recovery from lacquered wood.

and lacquered wood results in Figure 7.42. MS2 recovery and detection were inconclusive for the glass applications (Figure 7.43). The IVDS glass analysis demonstrated a very small peak at 24 nm—the size of MS2 measured with the IVDS. This
peak was not distinguishable from background and could not be labeled a positive
detection. The other samples showed distinguishable peaks above background.

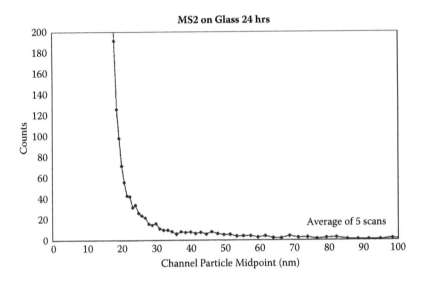

FIGURE 7.43   MS2 recovery from glass.

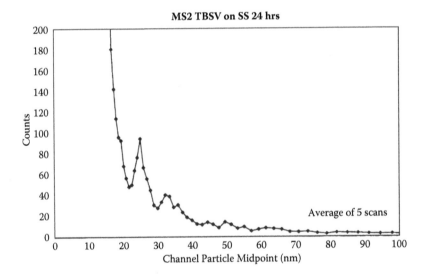

FIGURE 7.44   TBSV recovery from stainless steel.

TBSV was recovered and detected at reduced levels from the stainless steel (Figure 7.44) and glass (Figure 7.45) samples. The IVDS showed distinguishable peaks above background.

7.7.5   DISCUSSION

This simple demonstration successfully showed how simple it is to collect a virus from a surface and analyze its size and concentration. Both the MS2 and TBSV are

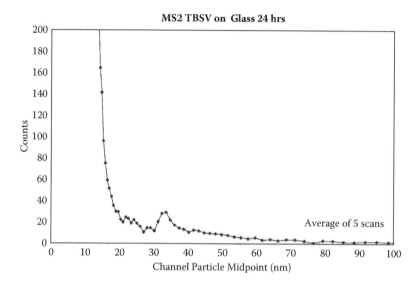

FIGURE 7.45   TBSV recovery from glass.

robust and could be expected to survive overnight on many surfaces. This demonstration indicates that this procedure can be followed for any virus and an effective dose response curve determined for a variety of variables such as temperature, pH, disinfection technique, routine cleaning method, etc.

Other non-enveloped viruses were examined using the ES-DMA-CPC combination. In addition to counting the viruses and separating multiple viruses in the same sample, Thomas demonstrated the separation of identical virus particles altered by packaged RNA. The difference between the two samples was less than 3 nm (Thomas 2004).

7.8   DETECTION OF VIRUSES IN BARK BEETLES

7.8.1   Introduction

Chapter 10 discusses viruses in honeybees. Viruses occur in other insects such as mosquitoes, ants, and beetles. The bark beetle infestation in the national forests has become an important issue. Like honeybee viruses, bark beetles are causing serious economic damage. For this reason we examined adult and larval bark beetles.

7.8.2   Methods

Two samples of mountain pine beetles were examined for viruses.

7.8.2.1   Adults (1.4g)
- Beetles were ground dry with mortar and pestle.
- Reverse osmosis water (160 mL) was added to remove ground mixture from mortar.

- Mixture was processed in blender for 60 seconds.
- Mixture was gravity filtered through cheesecloth (this took 15 minutes).
- Three 30 mL samples were prepared for centrifuge
- Samples were centrifuged for 60 minutes at 20,000g.
- Ultrafiltration through 500 MDa hollow fiber filters was performed.
- Sample size was reduced to 3 mL.

7.8.2.2 Larvae (4.2g)

- Larvae material was mixed with 160 mL reverse osmosis water.
- Mixture was processed in blender for 60 seconds.
- Mixture was gravity filtered through cheesecloth (this took 15 minutes).
- Three 30 mL samples were prepared for centrifuge.
- Samples were centrifuged for 60 minutes at 20,000g.
- Supernatant was gravity filtered through a 20 μm Whatman 41 paper filter as preparation for ultrafiltration.
- Ultrafiltration through 500 MDa hollow fiber filters was performed.
- Sample size was reduced to 2.5 mL.

7.8.2.3 Preparation for IVDS Processing

- Samples were diluted 1:10 with ammonium acetate buffer.
- Samples were filtered through a 20 μm Whatman 41 paper filter.

7.8.3 RESULTS

Figure 7.46 shows a 1034 count at 23.3 nm in the adult sample. Figure 7.47 shows a 3597 count at 25 nm. It would appear that the adult and larval beetles carry different viruses. The identification of the viruses could be achieved easily by other means (Wick 2011). The purpose of this experiment was to apply IVDS detection methods to insects.

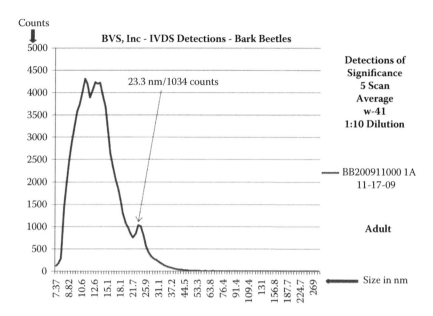

FIGURE 7.46 IVDS detected unknown virus at 23.3 nm in adult bark beetle.

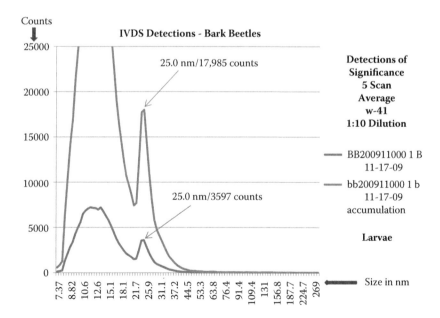

FIGURE 7.47 IVDS 3597 count at 25 nm in larvae bark beetles.

8 Detection of Other Nanoparticles

8.1 INTRODUCTION

In addition to studying viruses, the IVDS allowed us to examine other nanoparticles as potential standards or as interesting structures The most useful particles were obtained from the National Institute for Standards and Technology (NIST) and from biological supply sources. These standards were used to verify the correct functioning of IVDS and confirm the correct size detection.

Non-virus particles examined included gold particle standards, color standards, and polystyrene beads. Particles produced by precipitation were found to frequently have wide distributions with mean average sizes approximating the standards represented. We determined that the MS2 virus was the best size standard because of its consistency and narrow particle distribution suited to the 4-nm resolution of IVDS.

IVDS was able to detect many proteins, salts, and other such particles found in biological preparations. Most were determined to have been separated from the virus particles of interest and were not examined further. It is well within the capability of IVDS to investigate a wide range of non-virus particles and characterize them for various uses.

Various other instruments have been used to examine nanoparticles. One such instrument was made by Malvern and a short comparison with IVDS was considered useful because it revealed differences in technologies and confirmed the utility and accuracy of IVDS for examining particles (Wick et al. 2010c).

8.2 PARTICLE SIZE RESULTS: COMPARISON OF MALVERN NANO ZS AND IVDS

The Malvern Nano ZS measures zeta potentials or electrophoretic mobilities of fine particle systems including bacteria and viruses in solutions. The characterization of the surfaces of microorganisms is achieved by analyzing the surface charge and layer effects. The Nano ZS can assess the stabilities of biological organisms in suspension and emulsions and explore metabolic changes in organisms caused by matrix or environmental variables.

Various parameters such as pH, ion interactions, solution composition, and other *in vitro* parameters can affect the surface charge of a specimen. This non-disruptive analysis uses small sample sizes and leaves microorganisms intact for further sample analyses. The system also uses a non-invasive back scatter (NIBS) technology for particle analysis and size distributions.

8.2.1 METHOD

Particle size results generated by the Malvern instrument and the IVDS were compared. Polystyrene latex (PSL) particles from Bangs Laboratories in 20, 30, 70, and 90 nm sizes along with 30-plus-90 nm and 30-plus-130 nm mixed samples were measured.

8.2.2 RESULTS

The Malvern instrument analyzed the 20 nm, 30 nm, 70 nm, and 90 nm particles, and the scans are shown in Figures 8.1, 8.3, 8.5, and 8.7, respectively. The zeta (Z-Avg) values in the tables in the figures indicate diameter results in nanometers. The IVDS measured the same-sized samples and the scan results are shown in Figures 8.2, 8.4, 8.6, and 8.8, respectively.

The IVDS and the Malvern instrument were in close agreement for the major peaks of the PSL standards. The IVDS instrument shows a smaller extra peak in the 70- and 90-nm standards that does not appear in the Malvern results.

To examine the abilities of both instruments to differentiate several peaks, two mixed standards were prepared. A 30-plus-90 nm sample was mixed in a 1:3 ratio by volume (30 nm = 10). The Malvern results are shown in Figure 8.9. The Z-Avg was 79.23 nm with only one peak shown. The IVDS detected both standards as shown in Figure 8.10. Conversations with Malvern technical support indicated that the instrument needed an approximate 3× size difference to differentiate peaks.

A 30-plus-130-nm sample was mixed in a 10:1 ratio by volume to determine whether the Malvern device could differentiate the peaks. Figure 8.11 shows the separate peaks of the intensities of the standards. Figure 8.12 shows the size distribution by volume percent. The full-scale scan results of IVDS analysis are shown in Figure 8.13 and an expanded-scale scan in Figure 8.14. The IVDS was able to distinguish both PSL size standards.

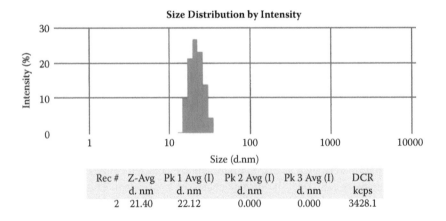

Rec #	Z-Avg d. nm	Pk 1 Avg (I) d. nm	Pk 2 Avg (I) d. nm	Pk 3 Avg (I) d. nm	DCR kcps
2	21.40	22.12	0.000	0.000	3428.1

FIGURE 8.1   Malvern results for 20 nm polystyrene latex standard.

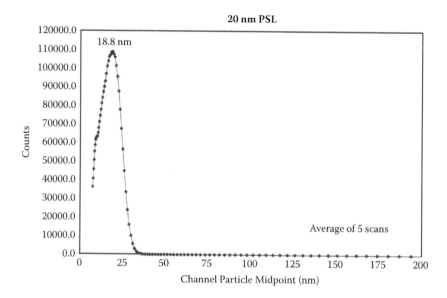

FIGURE 8.2 IVDS results for 20-nm polystyrene latex standard.

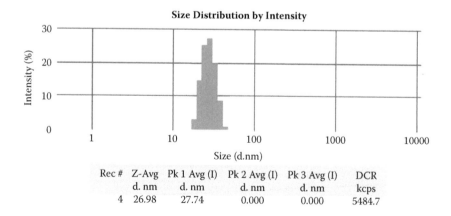

FIGURE 8.3 Malvern results for 30-nm polystyrene latex standard.

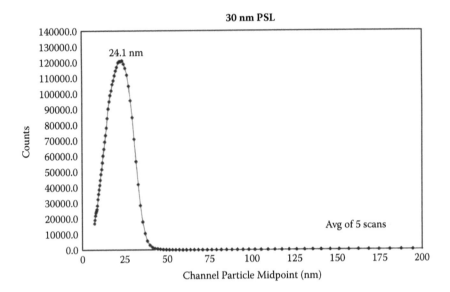

FIGURE 8.4   IVDS results for 30 nm polystyrene latex standard.

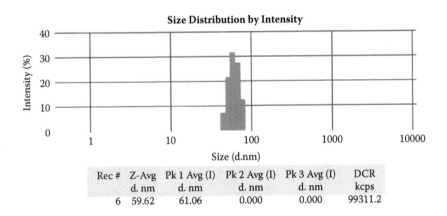

FIGURE 8.5   Malvern results for 70 nm polystyrene latex standard.

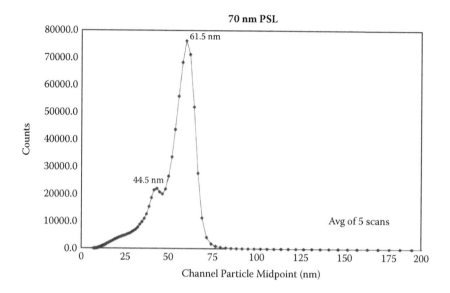

FIGURE 8.6   IVDS results for 70 nm polystyrene latex standard.

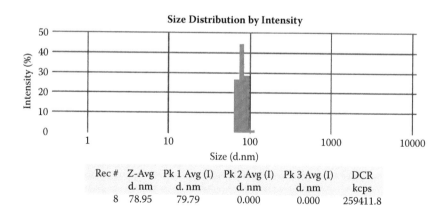

FIGURE 8.7   Malvern results for 90 nm polystyrene latex standard.

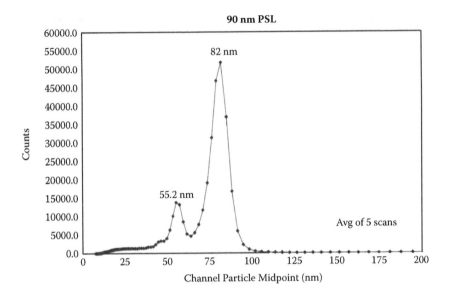

FIGURE 8.8   IVDS results for 90 nm polystyrene latex standard.

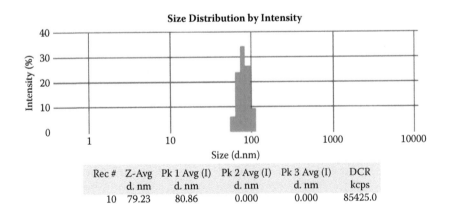

FIGURE 8.9   Malvern results for 30-plus-90 nm polystyrene latex standard.

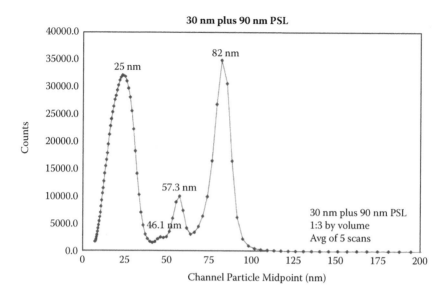

FIGURE 8.10   IVDS results for 30-plus-90 nm polystyrene latex standard.

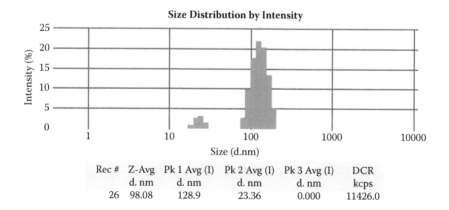

FIGURE 8.11   Malvern intensity results for mixed 30-plus-130 nm polystyrene latex standard.

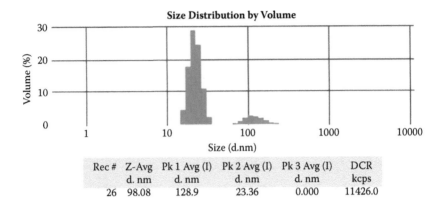

FIGURE 8.12 Malvern size distribution results for mixed 30-plus-130 nm polystyrene latex standards.

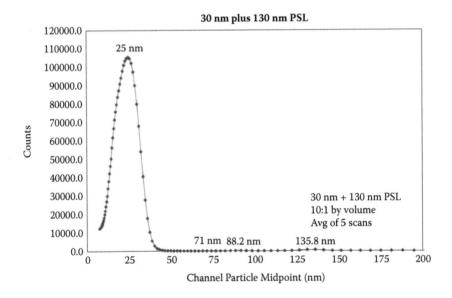

FIGURE 8.13 IVDS full-scale scan of 30-plus-130 nm polystyrene latex standard.

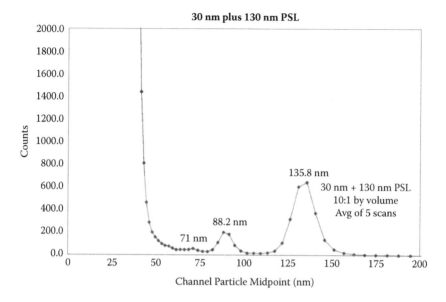

FIGURE 8.14   IVDS expanded-scale scan of 30-plus-130 nm polystyrene latex standard.

9 Direct Counting of Pili

9.1 INTRODUCTION

This chapter delineates the direct counting of "hair-like" structures (pili) specific for Gram-positive bacteria using IVDS and confirmed by mass spectrometry. Indications show that these structures are intact after removal from cells and are sufficiently different from species to species of bacteria. Thus, they can indicate bacteria type if not specific identification. Detection of pili represents a new approach to bacteria detection and identification. This report documents the detection of the bacterial structures using the physical nanometer counting methodology of the IVDS and electrospray ionization mass spectrometry techniques (Wick 2010a).

9.2 DETECTING PILI

This report explores the direct counting of hair-like structures specific for Gram-positive bacteria, as shown in Figure 9.1. Pili are found on many other classes of bacteria and led to the interesting possibility of devising a direct method for analysis by counting their numbers and determining their concentrations. These structures remain intact after removal from cells and are sufficiently different from species to species of bacteria to indicate bacteria type and possibly identification (St. Geme et al. 1996; Fader et al. 1982).

Their detection would represent a new approach to bacteria identification. This chapter documents the detection of the bacterial structures using the physical nanometer counting methodology combining the IVDS and electrospray ionization mass spectrometry. Bacteria and viruses are completely different types of microorganisms. Bacteria are a magnitude larger in size than viruses and have a specific classification scheme. Bacteria have cell walls, are organized into cellular components, and generally are considered self-sustaining organisms. In comparison, viruses are sometimes not even considered to be alive and their life cycles are completely different from those of bacteria. These physical and physiological differences of bacteria and viruses have hindered the development of a universal detector capable of simultaneous detection and characterization of both types of organisms.

The hair-like structures found on all classes of bacteria make possible the use of a virus detector based on physical principles to detect and quantify the unique pili structures that behave like particles when separated from a bacterium. Electrospray ionization mass spectrometry (ESI MS) has proven an effective technique for accurately identifying various microorganisms based on their proteomic or genomic contents (Yates 1998; Thomas et al. 1998; Krishnamurthy et al. 2000). Bacterial pili consist mainly of protein molecules, making ESI MS a suitable technique for

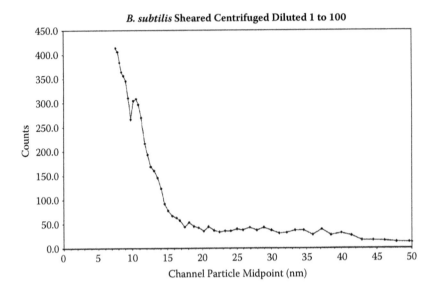

FIGURE 9.1   Bacterial pili sheared from *Bacillus subtilis* (three-scan average, BS2).

examining them and acquiring complementary information to add to data obtained from IVDS analysis. IVDS can differentiate particles from 5 to 950 nm in size.

Bacteria are generally 0.5 to 1 micron wide and 2 to 3 microns long—well beyond the measurement ability of IVDS. There are, however, interesting features on the surfaces of bacteria that fall in the proper size range for IVDS to characterize. Gram-negative bacteria, named for their inability to retain a crystal violet–iodine complex stain, have rigid surface appendages called pili. These hair-like structures, ~7 nm in diameter and varying in length up to 25 nm for longer flagellates (Eisenberg 1999), may be attached to the surfaces of bacteria. The pili are composed of structural protein subunits called pilins. Some pili have only single structural protein units; others are more complex and have several units.

These pili are characterized by a precise helical arrangement of one or more types of proteins and have different lengths for different bacteria. Choudhury et al. (1999) and others have studied crystal complexes associated with pilin subunits. Cell lysis breaks cells into components. Lysis can be achieved by changes in pH, temperature, ultrasonication treatment, or the addition of chemical reagents. The pili can then been treated like other nanometer-sized particles: separated and counted. Different bacteria exhibit different pili proteins, as expected. Evidence of such differences suggests that IVDS is capable of analyzing these virus-sized bacteria components and detecting bacteria.

Recent advances in the field of proteomics involving comprehensive mapping of the expressed proteins in organisms provide promising strategies for characterizing bacteria based on their protein biomarkers (Fleischmann et al. 1995; Ushinsky et al. 1997; Yates 1998; Thomas et al. 1998; Krishnamurthy et al. 2000). Mass spectrometry (MS) is widely used in the proteomic field for the characterization of various microorganisms by identifying as many protein components as possible biomarkers

(Yates 1998) or identifying one or several protein component(s) in a complex mixture (Thomas et al. 1998; Krishnamurthy et al. 2000). Various MS techniques have been used more extensively to characterize bacteria than for examining viruses (Thomas et al. 1998; Krishnamurthy et al. 2000; Dai et al. 1999).

Many factors contributed to the large volume of studies of bacteria, for example, the significant number of protein biomarkers encoded in bacteria and the thermal stabilities of bacterial proteins upon introduction into the MS ionization source (Tong et al. 1999; Karty et al. 1998; Bothner et al. 1999; Valegard et al. 1990).

MS has been used also to study the filamentous protein structures found in the pili of bacterial cells (Tito et al. 2000; Thomas et al. 1998; Cargile et al. 2001). Most studies focused on investigating the post-translation modifications of the pili proteins that play important roles in carrying diseases from various microorganisms, mainly *Neisseria meningitides* (Stimson et al. 1996). However, no attempts to investigate the pili proteins in *Bacillus subtilis* using MS or IVDS have been reported in the literature. The existing studies address the characterization of the pili proteins of *B. subtilis* samples processed via various techniques. MS serves as a complementary technique to IVDS if accurate and discriminatory characteristics of biomolecules must be revealed.

9.3 TESTING PROCEDURES

9.3.1 IVDS

A strain of *B. subtilis* (ATCC 15561) was prepared according to ordinary protocols. Briefly, the bacteria samples were grown on nutrient agar plates at 30°C for 26 hours. The bacterial cells were collected by scraping them into distilled water and pelleting them by centrifugation at 12,000g for 10 minutes. The bacterial samples were washed four times in distilled water by centrifugation, then resuspended in distilled water.

A 0.5-mL aliquot of the stock solution was passed in and out of a 22-gauge needle to break the pili from the surface of the bacteria. The shearing action of repeated passages through the needle has been shown to remove pili from bacteria (Baron et al. 2001). Each sample was centrifuged at 12,000 rpm (10,000g) for 25 minutes to pelletize the large bacterial cells. The supernatant (designated BS2) was collected and diluted 1:100 with 20 mM ammonium acetate for examination in the IVDS. Figure 9.1 shows the peak between 10 and 12 nm associated with bacterial pili.

The BS2 sample was filtered by the ultrafiltration subsystem of the IVDS using a 100 KDa filter (for details of the IVDS subsystems, see Chapters 1 and 2). The filtration system is designed to remove any materials with molecular weights smaller than the filtration system setting ((100 KDa in this case), e.g., growth media, salt molecules, and proteins from the solution (Wick et al. 1999) and leave a concentrated pili solution. The filtered sample (designated BS3) was then diluted 1:10 with 20 mM ammonium acetate before analysis with the IVDS. Figure 9.2 shows the ultrafiltered bacterial sample with a peak at ~12 nm.

The original sheared sample was also ultrafiltered before dilution using a 100 KDa filter and analyzed by the IVDS. After filtration, the sample (designated BS1) was diluted 1:10 with 20 mM ammonium acetate before IVDS analysis. Figure 9.3 shows the ultrafiltered neat bacterial sample with the peak at ~12 nm.

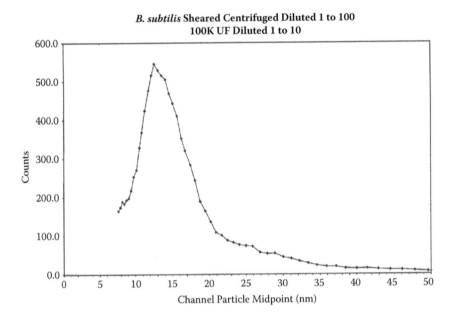

FIGURE 9.2 Ultrafiltered bacterial pili (five-scan average, BS3).

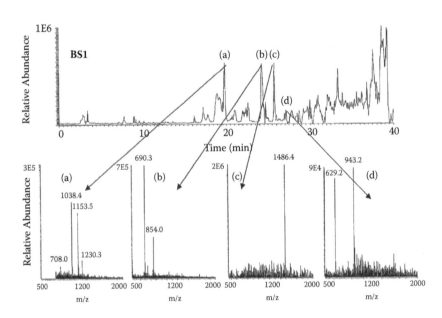

FIGURE 9.3 *Bacillus subtilis* sample (BS1) sheared without further filtration. Upper graph represents total ion chromatogram of LC MS analysis. Lower graphs are mass spectra of dominant peaks observed in replicate runs, N = 3.

9.3.2 MASS SPECTROMETRY

9.3.2.1 Electrospray Ionization Mass Spectrometry (ESI MS) Parameters

An ion trap mass spectrometer (LCQ-Deca, Thermo Finnigan USA) equipped with an ESI source was used. The mass spectrometer was operated under the control of the X-Caliber program with a manual deconvolution algorithm. Spectra were collected in the positive ion mode. Three microscans were used with a maximum ion injection time of 200 ms. The ESI spray voltage was maintained at 4 kV. The capillary voltage was maintained at 23 V. The temperature of the ion transport tube was 190°C. The mass spectrometer was calibrated to achieve a plus–minus Dalton resolution using a mixture of cytochrome C, bovine serum albumin (BSA), insulin, and myoglobin proteins.

9.3.2.2 Automated Deconvolution Algorithm Analysis

An automated deconvolution algorithm was developed to provide a filtered mass list instead of a conventional peak-at-every-mass output for the mass spectrometry proteomics (MSP) analysis of bacterial extracts. Briefly, this software deconvolutes the bacterial protein masses through analysis of a raw mass spectral file. Mass range, isotope peak width, signal-to-noise threshold, and maximum returned peaks based on user input parameters were selected before the start of deconvolution.

The software identified LC peaks and deconvoluted their corresponding average mass spectra to generate a list of masses. The deconvolution process takes 10 minutes for a 60-minute MSP analysis. The deconvolution step interfaces with relational database management software to update the local database with experimental bacterial protein mass data.

9.3.3 RESULTS

9.3.3.1 IVDS

The sheared *Bacillus subtilis* sample shown in Figure 9.1 displays the pili protein at ~10 to 12 nm next to a very large peak resulting from cellular and growth media contamination. The ultrafiltration molecular weight cutoff (MWCO) of 100 KDa allowed the removal of growth media and salts while retaining the pili fragments. After ultrafiltration, the peak showing the pili in solution was more pronounced and not masked by other contaminants.

To determine whether the 1:100 dilution preceding ultrafiltration and IVDS analysis affected the pili, a neat sample was ultrafiltered after shearing and centrifuging. IVDS results showed no significant differences from the diluted and ultrafiltered sample shown in Figure 9.2.

9.3.3.2 Mass Spectrometry

The unsurpassed sensitivity of MSP analysis to biomolecules makes the technique ideal for determining the molecular weights of intact proteins and identifying them. The extensive genomic and proteomic databases provided relevant information for identifying the biomolecules present in the analyzed samples in this study. Application of liquid chromatography combined with mass spectrometry (LC MS) to the characterization of biomolecules originating from various parts of bacterial cells has been successful in many studies (Yates 1998; Thomas et al. 1998; Krishnamurthy et al. 2000; Dai et al. 1999; Jensen et al. 1999).

In our study, the *Bacillus subtilis* samples underwent different sample processing approaches. Some samples were sheared and centrifuged without further filtration; they were labeled BS1. The other *B. subtilis* samples were exposed to similar sample processing, then subjected to centrifugation, dilution, and finally MWCO using a 100 KDa membrane to remove cellular debris and large particulates; these samples were labeled BS2 and BS3.

Figure 9.4 shows the LC MS results from the BS1 sample. The upper section represents the total ion chromatogram (TIC) of BS1 and the lower section shows mass spectra of the dominant proteins detected by the ion trap mass spectrometer. The peaks appearing toward the end of the TIC resulted from the large amount of buffer in the sample; this is reflected by the non-symmetrical peak shape characteristic of a mixture of non-proteinaceous molecules.

Upon examining the mass list generated from the LC MS analysis of BS1, we determined that the mass spectra represent proteins characteristic of the pili and

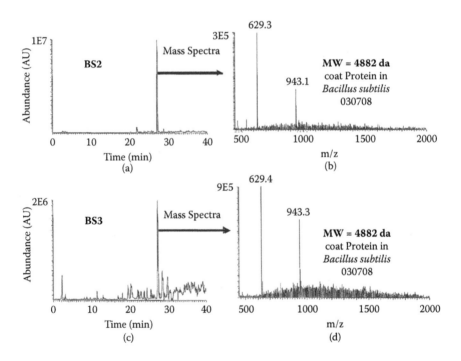

FIGURE 9.4 Effect of concentration on reproducibility of LC MS analysis of pili extracts of *Bacillus subtilis*. (a) Total ion chromatogram of *B. subtilis* sample (BS2) sheared, centrifuged, and filtered using 100 KDa ultrafiltration membrane. (b) Mass spectrum of most intense peak from TIC of BS2. (c) Total ion chromatogram of *B. subtilis* sample (BS3) sheared, centrifuged, and filtered using 100 KDa ultrafiltration membrane and 100-fold dilution over BS2. (d) Mass spectrum of most intense peak from TIC of BS3.

flagella parts of bacterial cells. A comparison of the experimentally deconvoluted protein molecular weights with weights shown in the Swissprot public database showed a match of the deconvoluted proteins with the outer coats and the extracellular proteins associated with pili and flagella parts of bacterial cells (Lei et al. 1998).

The LC MS analyses of the sheared and filtered *B. subtilis* samples BS2 and BS3 showed significant resemblances of their TIC plots and the dominant bacterial proteins identified. It is interesting to note that the TIC plots (Figure 9.4) are characterized by a dominant peak that has the same retention time observed in the LC MS analyses of these bacterial extracts. This peak is indicative of a significant expression of the pili protein compared to the other observed, but unnamed, proteins. The mass spectra of this dominant peak revealed that the same protein was found in both samples with better signal-to-noise ratio observed in BS2. The difference in signal-to-noise ratio could arise from the fact that BS3 was diluted 100-fold more than BS2.

The peak was identified as a coat protein of *B. subtilis*, which is an indication of the effectiveness of the sample processing approach in isolating the desired parts of bacterial cells—pili in this case. It is also worth mentioning that this coat protein was observed in the BS1 sample but with lower peak intensity than observed in the TIC plots of the BS2 and BS3 samples.

The LC MS data were further examined to determine reproducibility and the effects of sample processing on the analyses of *B. subtilis* extracts. Figure 9.5 shows the reproducibility of the molecular weights of the experimentally deconvoluted proteins generated from the LC MS analyses.

The masses generated for these samples were then averaged from three replicate LC MS analyses per sample. The statistical correlation study showed that a higher

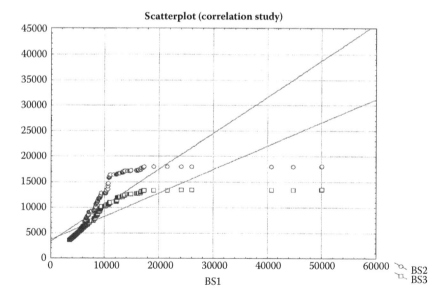

FIGURE 9.5 Molecular weight correlation of BS1 versus BS2 and BS3. Only BS2 and BS3 were exposed to ultrafiltration with molecular weight cutoff of 100 KDa membrane.

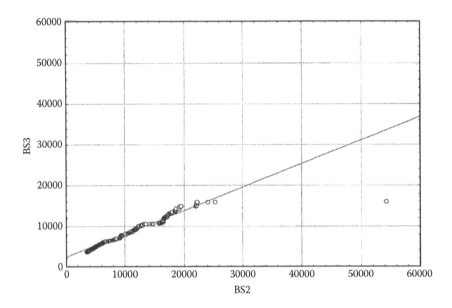

FIGURE 9.6   Molecular weight correlation of protein masses generated from *Bacillus subtilis* samples BS2 and BS3. Both samples underwent similar sample processing procedures; the latter was diluted 100-fold more than the former sample.

number of common masses were observed between BS2 and BS3 than with BS1. Figure 9.5 also reflects the relevance of comparing the mass lists generated from differently processed bacterial extract samples.

The increase of correlation factors among samples indicates a resemblance in the sample processing approach as evident in this case. Samples BS2 and BS3 were processed under similar conditions with only a difference in dilution as compared to BS1. This observation was also supported by a comparison of the deconvoluted mass lists generated for BS2 and BS3 indicating that more than 99% of the masses matched with each other (Figure 9.6).

The correlation factor of 99% is observed between BS2 and BS3, as compared with 80 to 81% match with BS 1. Overall, the LC MS served as a technique to confirm data obtained from IVDS analysis. The similarity between the IVDS and LC MS results from *B. subtilis* samples was observed. The similarity in comparing the IVDS data among the *B. subtilis* extract samples was also observed within their corresponding LC MS analyses.

9.4   CONCLUSIONS

The IVDS system is well suited for rapid physical measurements of the submicron pili of bacteria after mechanical removal. The pili were shown to be from *Bacillus subtilis* through mass spectrometry analysis. The collected LC MS data of the bacterial samples showed that most detected proteins belonged to the pili regions in cells. The ability of LC MS to decipher the identities of the bacterial proteins and track their positions in the bacterial cells qualified the technique as complementary and confirmatory to IVDS identification of bacterial pili based on their sizes and concentrations.

The higher sensitivity of the LC MS technique as compared to that of IVDS means that pili samples may require further preparation before their analysis by LC MS, as indicated by the interference peaks that appeared in some of the MS spectra (Figure 9.4). It should be possible with further experimentation to correlate the rapid identification of bacterial groups with the straightforward counting methodology of the IVDS and their proteomic identification using LC MS.

10 Detecting Viruses in Honeybees

10.1   INTRODUCTION AND FORMATION OF BVS, INC.

The new IVDS technology attracted interest from organizations outside the US Army. Many groups were interested in focusing this technology onto their problems, for example, in the areas of plant disease control, animal monitoring, and food source inspections. Two problems, the plights of white nose bats and honeybees were examined further. The honeybees constituted an attractive first environmental sample because of the apparent ease of collecting samples and the possibilities of multiple viruses. The IVDS was designed to determine many viruses from a single sample and a field application was needed to demonstrate efficacy.

Early collaboration with other government groups also focused on insects and the needs to collect and measure viruses of interest. Thus, when a beekeeper approached one of these collaborative groups, the group contacted the US Army's Edgewood Chemical and Biological Center (ECBC) and the honeybee connection was made.

The first samples that arrived at ECBC were in small bags containing lots of bees. The bags arrived on my desk with some of the bees still buzzing and moving around. Because beekeeping was not our specific area of experience, we had to determine methods for processing the bees for analysis. This was a suitable challenge because our detector was expected to operate in the field and be omnipotent.

Processing the honeybees turned out to be a relatively simple matter of grinding them up in water, filtering the solution, and running samples through the regular IVDS process. The initial grinding process took a little practice because we had no experience with these types of samples. The first devices were simple off-the-shelf coffee grinders. We found that a blender solved the early routine sampling and initial processing issues. We were able to detect viruses quickly and then related the information to entomologists who asked for our help. Results connected a virus and a fungus that reduced honeybee health (Bromenshenk et al. 2010).

BVS, Inc., in collaboration with the beekeeping industry and other parties brought the IVDS to its facility in Missoula, Montana, and obtained a patent license agreement from the US Army for this purpose. The objective was to use the new technology to examine virus issues related to colony collapse disorder and analyze honeybee health. BVS, Inc. also executed a research agreement with the Army laboratory at ECBC.

Delays at the start of operations were overcome and bee screening commenced in May 2008. In cooperation with Montana beekeepers and Bee Alert Technologies of Missoula, samples were collected and logged starting in January 2008. All backlogged samples (almost 800) were processed in a unified effort intended to provide a more complete picture of bee health as a management tool.

BVS assembled a database for detecting viruses in bees. The database information is offered to beekeepers in the forms of reports and presentations. The database is unique in that it provides information that was not previously available to beekeepers. The IVDS data reflects the sizes of detected viruses in nanometers (nm). While this is a very accurate and consistent method, we still do not have names of all individual viruses and size correlation data. However, the system has detected viruses that appear relative to bee health. Our next step is to continue to work with the Army, Bee Alert, and others to correlate detected sizes with virus names.

Viruses have long concerned beekeepers as possible and probable causes of bee deaths, increased expenses, and management uncertainties. Traditional methods for virus management often are unreliable due to the limited knowledge of bee viruses and the difficulty in determining whether viruses are present in colonies. The relationship between a virus and a fungus was established in a widely read paper (Bromenshenk et al. 2010).

Sample preparation consists of placing bees in a blender in distilled water, blending, coarse filtering, and centrifuging to remove large materials. Ultrafiltration with 500,000 Da hollow-fiber filters was used to remove background materials and small particles (salts). The resultant solution isolates the viruses in suspension. Each resultant sample is converted into an electrically charged aerosol by the IVDS instrument, taking advantage of the physical movements of nanoparticles in charged atmospheres. Each particle size is separated and recorded; this requires about 5 minutes per sample and no reagents. Total sample preparation time is about 2 hours but using several sample preparation stations processing several samples in parallel allows many samples to be prepared and processed daily.

The IVDS process allows rapid screening for all intact viruses. Each virus has peak of a specific nanometer size that is recorded. The height of a peak indicates the concentration of the active virus. Honeybee viruses have different and specific sizes that enable us to distinguish multiple viruses in a single sample. We use more traditional molecular methods (e.g., proteomic mass spectrometry, polymerase chain reaction, and microarrays) to identify each virus of a given size. Subsequent IVDS peak detections of a specific size are indexed with a specific identified name. Periodic verifications are used to assure accuracy and consistency.

10.1.1 EARLY OBJECTIVES

The first objective was to demonstrate and promote the use of the IVDS technology via field sampling, direct reporting to beekeepers, meeting with beekeepers in their apiaries, and reporting at the Montana, regional, and national beekeeping association meetings. Equally important was beginning to use IVDS analysis of honeybees to promote honeybee health by monitoring virus loading as a function of conditions such as environmental factors (temperature, relative humidity), feeding and management practices, and impacts associated with transport and relocation. Many variables affect honeybee health and our objective was to apply this new technology to the beekeeping industry. Several steps were identified early:

- Demonstrate the ability of IVDS to detect viruses in a single honeybee
- Associate the many virus peaks with names of viruses (Rosetta Stone project)
- Assist in assessments of honeybee nutrition and health
- Evaluate the use of essential oils applied to honeybee health

10.1.2   EQUIPMENT AND METHODS

The basic setup required for the IVDS laboratory in Montana initially consisted of:

- Bee grinding and coarse filtration
- Centrifuge
- Ultrafiltration
- IVDS

The following methods were designed:

- Bees were collected at the direction of Bee Alert Technologies and delivered to BVS for processing.
- Each sample was to contain 6g of bees for standardization. However, if a sample did not have enough bees for the 6g, the entire sample was processed with appropriate notation. Each sample was blended with 100 mL of reverse osmosis (RO) water and coarse filtered through single-layer cheesecloth.
- Next was centrifugation of 30 mL of sample for 60 minutes at 19,000 × gravity (g).
- Supernatant was to be recovered and ultrafiltered through a 100,000 Da hollow fiber filtration system and a 300 mL RO wash and reduced to ~3 mL.
- IVDS preparation consisted of 1:10 and 1:100 dilutions with ammonium acetate (AA).
- Five IVDS scans were to be averaged and saved in database.
- Finally, creation of charts and tables from data exported from IVDS.

10.1.3   RESULTS AND LESSONS LEARNED

Setting up the laboratory revealed limitations that were not anticipated in the original design. The first issue was the length of time required for the ultrafiltration stage to process a single sample. The initial correction was the addition of filtration stations. The addition of one station doubled output. Adding another system further increased production. The systems could all be scaled to accommodate the number of samples processed. We also replaced the filtration cartridges with higher-throughput cartridges tested by ECBC that enabled us to process more samples per filtration station per day.

The second issue concerned centrifuges. We discovered early that a pair of centrifuges could not handle the number of samples run per day even though we did not exceed their continuous ratings. We worked with the manufacturer (Themofisher). After replacing one centrifuge with three new ones and replacing the second machine with two more, we finally settled on a single centrifuge of higher capacity and rating that the manufactured determined was the right tool for the job. That solved the centrifuge problem. After a quick learning curve, the IVDS was found to work exactly as expected. The IVDS instrument was the easiest component to install; it worked out of the box and has been working consistently for many years. After resolution of the filtration and centrifuge issues, the production laboratory was in operation.

Screening for viruses generated considerable quantities of data, particularly in particle sizes ranging from 17.5 to 46.1 nm—the region of interest for particles found in honeybees. Particle detection in honeybees by the IVDS was considered virus detection based on sample preparation and the natures of viruses in solutions. The sample sets reported in this chapter were generated at the BVS laboratory in Missoula that examined over 8,500 honeybee samples.

All of the detection data cited in this section came from basic IVDS detection charts for all samples processed. Each chart showed individual peaks representing detections at specific sizes. Detection chart information was then compiled into tables and combined with other samples to create analytical charts. The most frequent virus detected was Sacbrood (32.2 nm); it was detected in nearly 45% of the samples.

Figure 10.1 shows the virus dynamics over a 3-year period. It indicates that some viruses have low occurrences and identifies the most frequent honeybee viruses and their numbers. Figure 10.2 shows no virus detection in a sample. Figure 10.3 shows the seasonal variations of honeybee viruses from late winter to late summer. These observations are consistent from year as seen in Figure 10.1. The seasonal

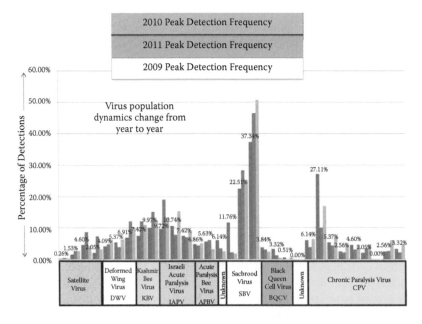

FIGURE 10.1 Honeybee virus population dynamics plotted as a function of percentage of detections for 3 years (2009, 2010, and 2011). The most common finding was Sacbrood.

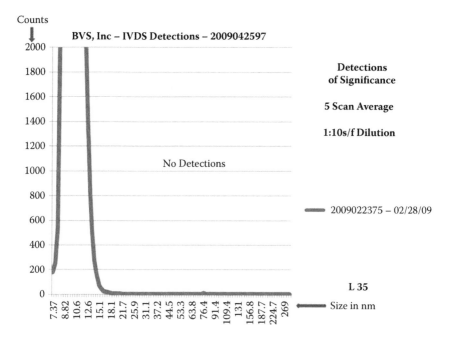

FIGURE 10.2 Typical IVDS honeybee analysis showing no viruses.

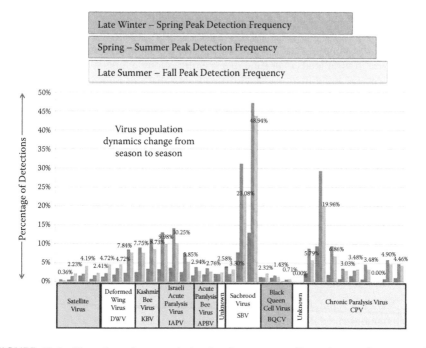

FIGURE 10.3 Honeybee virus population for three seasons (late winter to late summer) presented as percentages of detection. The most common viral occurrence for all seasons was Sacbrood.

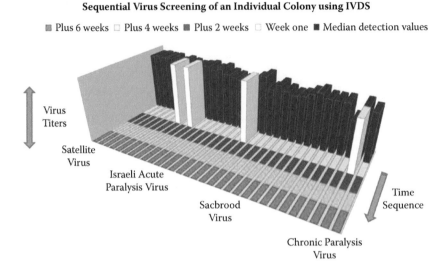

Sequential Virus Screening of an Individual Colony using IVDS

Plus 6 weeks ☐ Plus 4 weeks ■ Plus 2 weeks ☐ Week one ■ Median detection values

Virus Titers

Satellite Virus

Israeli Acute Paralysis Virus

Sacbrood Virus

Chronic Paralysis Virus

Time Sequence

FIGURE 10.4 Virus detection titers from single hive. Median counts are indicated in backdrop. Four viruses are indicated.

observations established the common or normalized honeybee virus populations and are useful for further comparisons. They also establish baselines for developing cause-and-effect curves for analyzing long-term effects.

Samples show that we now have the ability to follow and compare individual colonies to all the samples processed. Comparisons of this nature will show increases or decreases in viral loads over time in a specific colony or apiary. Reports can then integrate several combinations of viral load data that are relevant to maintaining honeybee health.

Figures 10.4 through 10.7 compare results of a colony over time. Data were obtained initially and then after 2, 4, and 6 weeks. Figure 10.5 (after 2 weeks) reveals noticeably lower counts and fewer peak detections. The virus information changes at 4 weeks (Figure 10.6) and again in 6 weeks (Figure 10.7).

Figures 10.8 through 10.10 depict dose response curves as a function of treatment and viral loading for a colony treated for Nosema. This curve, shown in three steps, shows a dramatic decrease in virus titer and much less virus diversity at the end of the treatment. It should be noted that the first result of treatment was a rapid increase in the virus load followed by a rapid decrease in viral loading at the end.

10.1.4 EARLY ACCOMPLISHMENTS

- IVDS is a rapid, low-cost approach to virus detection.
- It demonstrated detections of virus-sized particles.
- Peak identifications were verified by using proteomics and other identification methods.

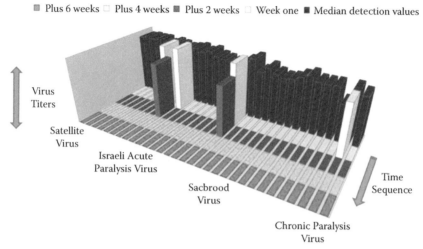

Sequential Virus Screening of an Individual Colony using IVDS

■ Plus 6 weeks □ Plus 4 weeks ■ Plus 2 weeks □ Week one ■ Median detection values

FIGURE 10.5 Continuation of observations from Figure 10.3 into second week. Notice absence of two viruses.

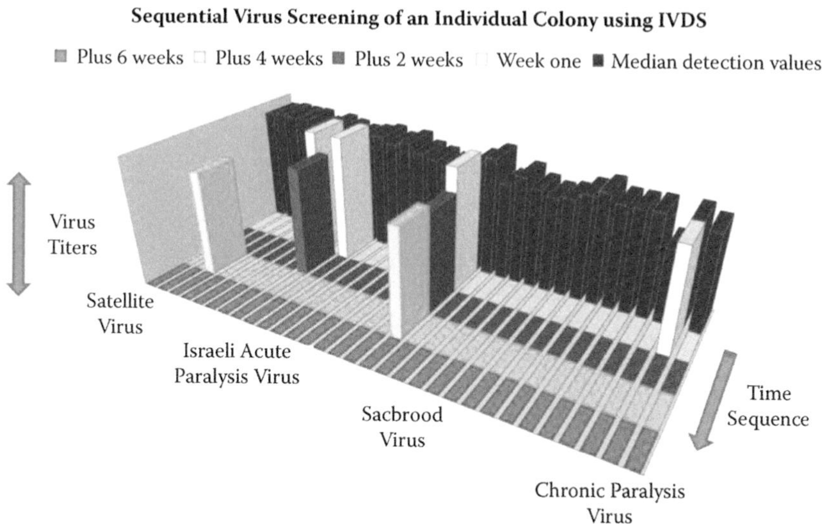

Sequential Virus Screening of an Individual Colony using IVDS

■ Plus 6 weeks □ Plus 4 weeks ■ Plus 2 weeks □ Week one ■ Median detection values

FIGURE 10.6 Observations continued from Figure 10.4. A new virus is added to the mixture and one (IAPV) exits.

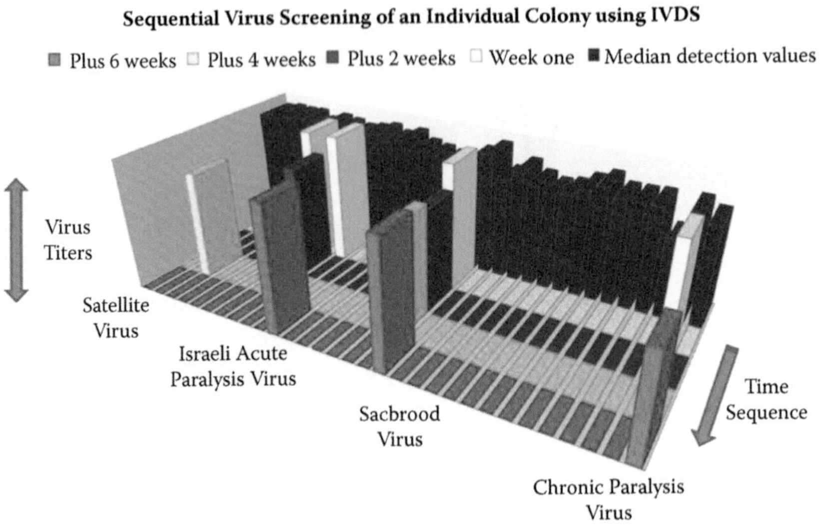

Sequential Virus Screening of an Individual Colony using IVDS

■ Plus 6 weeks ☐ Plus 4 weeks ■ Plus 2 weeks ☐ Week one ■ Median detection values

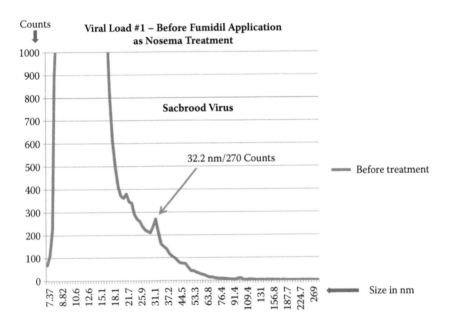

FIGURE 10.7 Observations continued from Figure 10.5 with additions of IAPV and CPV while the satellite virus exits. The common thread through all 6 weeks of observation was the Sacbrood virus.

FIGURE 10.8 First measurement of virus titer to assess treatment with Fumidil, a fungicide for Nosema. A peak is seen for Sacbrood.

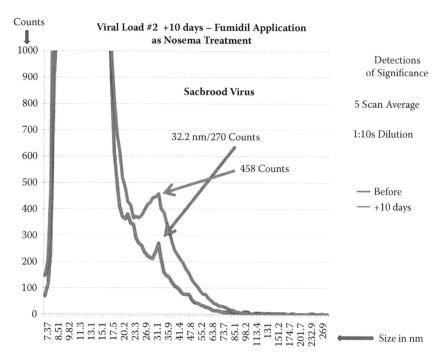

Counts

Viral Load #2 +10 days – Fumidil Application as Nosema Treatment

Sacbrood Virus

32.2 nm/270 Counts

458 Counts

Detections of Significance

5 Scan Average

1:10s Dilution

—— Before
—— +10 days

Size in nm

FIGURE 10.9 Second measurement following Fumidil treatment for Nosema taken 10 days later. Note increase in Sacbrood virus indicating reaction to treatment.

- Ongoing monitoring of samples for viral load changes can reflect the effects of bee management practices.
- This method of virus monitoring and detection does not require virus knowledge or experience.
- Peak detections of unknowns are analyzed by more traditional molecular methods.
- The large and growing data set enables broader statistical analysis for a better understanding of viral loads and distributions.

The data and the related charts produced from the IVDS were the results we sought and represent a success in the form of a useful tool for beekeepers. BVS would like to acknowledge the Montana Department of Agriculture and the Montana State Beekeeper Association for their help in making this type of analysis possible. A new way of looking at viral loads and virus management is available because of the vision and support of beekeepers and the industry.

Virus analysis has been difficult for beekeepers because of the costs and detection difficulties. We can now work to find solutions to improve viral management and aid beekeepers in maintaining healthy colonies. Correlation of viral data with information about transportation, feeding, local bee health issues, honey flow, wintering, and other beekeeping aspects will provide significant help in recognizing and managing viruses.

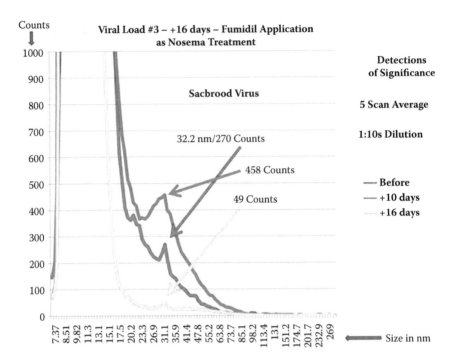

FIGURE 10.10 Same hive as in Figure 10.8 measured 6 days later shows decrease in Sacbrood and absence of other viruses indicating recovery from treatment and evident good health.

The next step was to perform the research and comparative work that will assign virus names to the peaks. After the initial success we were also ready to screen bees for viruses on a broader scale and recruit beekeepers to send more samples.

10.2   DETECTION OF VIRUSES FROM SINGLE HONEYBEE

An early test of the versatility of the IVDS was the examination of a single honeybee for viruses. A sample contained a honeybee queen and the results of testing were exceptional. Figure 10.11 shows a peak at 33.4 nm associated with the Black Queen Cell virus (BQCV). The association was determined by the Rosetta Stone project discussed in Section 10.3. Many other viruses were also associated, but not with this single honeybee.

10.3   ROSETTA STONE PROJECT

10.3.1   INTRODUCTION

The IVDS detects viruses and virus-like particles but does not identify them. It does, however, indicate the precise size of any virus detected and was found to be useful for dividing detected viruses into groups or more specifically groups associated with known viruses. In the case of the numerous honeybee viruses, a rigorous process was

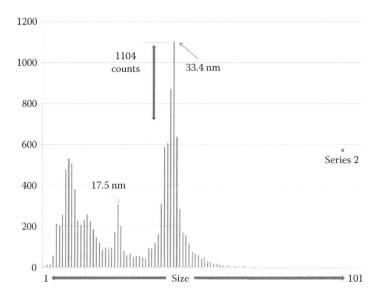

FIGURE 10.11 Virus count from single queen bee. The large count is consistent with Black Queen Cell virus.

followed for assigning virus names to the common peaks detected in honeybees. By using orthogonal methods to analyze the same samples or the same virus associations, it was possible to associate a virus size and name.

10.3.2 DETECTION COMPARATIVE ANALYSIS

The IVDS employs a detection method that is relatively simple, inexpensive, and rapid, even for emerging, mutated, and unknown viruses. This type of detector is indicated for processing large numbers of samples and screening samples that may or may not contain a virus of interest. By associating a virus size with a known virus, it becomes possible to look for viruses of a specific size as though they were known viruses. The name of a virus can be confirmed by other means at a later time.

Detection is a valuable step because the conformational determination may be an order of magnitude more expensive than the initial detection step. Frequently, as experience with the detection of a virus of a known size increases, the conformational step is unnecessary.

In the cases of the Honeybee Nutrition Study (Section 10.4) and the Essential Oils Study (Section 10.5), it was more important to detect numerous viruses and determine their relative concentrations than to identify them specifically. It is often not necessary to know the names of the viruses revealed on detection charts if the relative virus loading figures are important. Higher virus loads are usually alarming and low virus loading is acceptable or normal, as observed in many organisms. Honeybees were no exception. The intent is often to attempt to lower virus loads through interactions such as changes to the environment, nutrition, or other means.

10.3.3   Methods

The IVDS method was used to analyze samples that contained honeybee viruses. During the analysis, the samples were converted into electrically charged aerosols by the IVDS instrument. This step takes advantage of the movement physics of nanoparticles in charged atmospheres; each particle size is separated and recorded. The step requires about 5 minutes per sample and no reagents.

Each virus has a peak of a specific nanometer size. The peaks are recorded and the heights are indicative of the concentrations of active viruses. Each virus has a specific size that enables recognition of multiple viruses in a single sample.

Confirmatory analysis using molecular methods such as proteomics mass spectrometry, polymerase chain reaction (PCR), and microarrays were used to cross-analyze each virus. The results from using such techniques were compared to IVDS results for the same samples. Ideally this would have resulted in a one-to-one naming of each IVDS peak to a virus. However, variables sometimes presented challenges. For example, uncertainties of the genetic identification procedure were exacerbated by the very close genetic relationships among some viruses. Despite these close genetic characteristics and the resulting academic discussions about proper identification, the sizes of the particles remained constant for each virus. Subsequent IVDS peak detections of specific size were indexed with specific identified names. The IVDS emerged as a rapid and inexpensive method for monitoring honeybee viruses.

The labeling of IVDS peak detections was based on the correlation of virus size to confirming data from PCR, microarray, and proteomics analyses. The objective of this project was to establish virus names for specific peaks in IVDS data. While the IVDS technology provides rapid detection, it does not reliably compare virus sizes to names of the viruses. All the existing work generally covers ranges of sizes for each virus. The nature of the IVDS instrument allows very precise sizing for each virus based on its inherent physical properties and mass:charge ratios. Bee Alert Technologies, the University of San Francisco's DeRisi Laboratory, ECBC, the United States Department of Agriculture's (USDA's) Agricultural Research Service (ARS) Laboratory, and BVS collaborated with the beekeeping industry and with key researchers to utilize the labeling of IVDS data to develop a common terminology understood by all.

10.3.4   Results

Sample collection was designed for representative colony samples in cooperation with beekeepers, the USDA ARS, and Bee Alert. IVDS protocols were developed at ECBC. Reports and charts were developed at BVS. The data derived by correlating information developed by the three groups provided a basis for the IVDS to name viruses based on the following peak sizes:

- Deformed wing virus (DWV) at 21.7 nm
- Kashmir bee virus (KBV) at 22.5 nm
- Israeli acute paralysis virus (IAPV) at 25.0 nm

- Acute bee paralysis virus (ABPV) at 26.9 nm
- Sacbrood virus (SBV) at 32.2 nm
- Black Queen Cell virus (BQCV) at 33.4 nm

Two other peaks that appear in the IVDS data set do not show in results from other technologies, perhaps because no one looked for them or analysts had no primers for them. Further work may reveal the nature of these particles located at 17.5 to 18.8 nm. They may represent a single virus that many sources in the literature call satellite viruses or they may be fragments of some other entity. These virus-sized particles are consistent in IVDS results and may or may not be important. Observations indicate they are associated with chronic paralysis virus (CPV) or possibly another large virus. CPV is an ellipsoidal virus appearing at 30, 40, 55, and 65 nm. IVDS detected a broad peak ranging from 40 to 70 nm in some beekeeper samples although CPV has not been confirmed. The peak at 37.2 nm may be the bee X virus associated with CPV. More research including comparative analyses must be done.

Proteomic methods based on peptide sequence data generated from LC-ESI MS/MS analysis of protein digestion present a major advantage over other identification methods in that no prior knowledge about a sample is required. Although, it is obvious that taxa underrepresented in a database will not provide sufficiently high resolution to classify an unknown sample accurately to strain level, genus detection is possible.

When we compared data from various technologies, inconsistencies became apparent. Several explanations are possible but the inconsistencies are likely due to the different genetic sequences used or the systems were not sensitive to the related viruses present. In other words, specificity of detection was lost. A transmission electron micrograph (TEM) tends to show virus sizes ~30 nm. With other techniques, possible artifacts are created and maintaining proper structure integrity (size) becomes a major issue. The IVDS sized samples precisely at 17.1, 21.7, 26.9, 28.9, 32.2, 35.4, and 37.2 nm and revealed other sizes for other particles. When examining a sample containing multiple viruses, the diversity and similarity in size clearly complicate the task of visual separation by size.

10.3.5 Conclusion

IVDS is a broad-based virus detector that assigned names to six of the peaks associated with honeybee viruses. IVDS is much faster than the other techniques used to correlate names with peaks and is indicated as a method of choice for routine monitoring and characterization of virus loads in honeybees.

10.4 HONEYBEE NUTRITION STUDY

10.4.1 Introduction

Nutrition is essential to honeybee health. This simple statement leads to many questions. What types of nutrition are best? How should nutrition be delivered and in what quantities during different seasons? How does nutrition interact with various stress factors? Many variables affect the answers. One idea was to develop a dose response (or cause and effect) curve to assess nutrition factors. Another was to measure viral loads and diversity as functions of the variables because we know that viral loads change seasonally. Do the loads change because of nutrition practices? These questions led us to design several initial tests to evaluate nutritional factors.

10.4.2 OBJECTIVES

To assess honeybee nutritional issues adequately, we had to determine:

- Viral diversity changes over time caused by adding nutrition to the honeybee diet
- Viral titer change over time resulting from addition of nutrition to bee diet
- Nutrition levels required to effect changes
- Viruses most affected by nutritional changes
- Timing intervals for adding nutrition to produce highest impacts on viral loads

10.4.3 MATERIALS AND METHODS

Bees were collected and delivered to BVS for processing. Samples were prepared in a 50 mL solution containing 100 bees ground and mixed with 50 mL of deionized water. Samples were then frozen. Samples were then filtered through standard cheesecloth to remove non-soluble bee parts.

Each 30 mL of sample was centrifuged for 60 minutes at 20,000 × gravity (g). The supernatant was recovered and ultrafiltered through a 500,000-Da hollow fiber filtration system and a 250 mL reverse osmosis wash that reduced the sample to ~2 mL to concentrate the viruses. The sample solution was then prepared for IVDS analysis by 1:10 dilution using ammonium acetate as the salt for controlled conductivity. Next, each sample was filtered through W-41 20 μm paper or a 0.45 μm membrane filter to remove fatty and pollen residues that tend to float in a solution after centrifugation.

The IVDS calculated five-scan averages and saved them in a database. Charts and tables were then created from the data exported by the IVDS.

10.4.4 RESULTS AND DISCUSSION

As a part of the project, we addressed some of the procedures for collecting and processing samples to ensure data integrity. The quality of samples at various stages of collection, processing, storage, and filtration was a continuing concern, particularly with regard to viral intensity and diversity in samples and degradation of virus integrity. Storage conditions were monitored to ensure virus sample integrity at various temperatures and in several media. Sample integrity must be maintained to allow consistent data comparisons over time.

The best filter setup for honeybee application showing only minimal loss of titer (<10%) and no diversity loss was a 20 μm paper filter and syringe holder. The temperature storage and media applications for sample preservation showed no titer loss or diversity loss from vials holding frozen or fresh samples over time. The largest losses of apparent viron detection revealed by the IVDS were from partially processed samples stored in refrigeration; the loss was greater than 20% after 72 hours. However, earlier processing (within 24 hours) prevented apparent loss. The sample storage time preceding IVDS processing that showed no apparent loss was 1 week at 35°F. Processed samples ready for IVDS and frozen at –30°C did not reveal any degradation after a year.

The methods used for sample shipping, processing, and storage demonstrated integrity of the intact viruses processed by IVDS and increased the reliability of tracking viral titers and diversity over time using the IVDS. We investigated a time-of-day sample collection to determine variations in viral loads by analyzing the diversity frequency and comparative titer data on over 800 processed samples.

For the project assessing nutrition as a function of virus diversity and titers, the data set and sampling were scaled to accommodate the increased number of samples required to yield better statistical analysis. The initial data was consistent with previous work and showed titer and diversity changes over time. The changes had not been correlated fully with nutritional data and other variables such as such as varroa mite levels and relationships. These relations will be better understood as more data is collected and analyzed.

10.5   ESSENTIAL OILS STUDY

10.5.1   MATERIALS AND METHODS

The essential oil formulation for making 570 to 600 sample packets is shown in Table 10.1.

TABLE 10.1
Essential Oil Formulation

Ingredient	Quantity
Thymol crystals	15 pints
Eucalyptus oil	3 pints
Menthol crystals	1 pint
Lemon grass oil	1 pint
Tea tree oil	1 pint
Wintergreen oil	1 pint
Camphor oil	1 pint
Tylosil	2 bottles
Vegetable oil	25 pounds
Powdered sugar	25 pounds
Granulated sugar	50 pounds

10.5.1.1   Sample Collection

The samples for the Nosema–essential oil study were shipped via USPS flat-rate service in zip-lock bags. Sample weight was 6g or nearly 60 bees. The bee samples were liquefied in 300 mL of deionized water. The samples for viral load monitoring and stress-level association were shipped and prepared in the same manner.

10.5.1.2   Sample Processing

Each sample received by BVS was filtered through standard cheesecloth to remove non-soluble bee parts and adjusted to 30 mL by adding ammonia acetate. Each 30 mL sample was centrifuged for 60 minutes at 20,000 × g. The supernatant was recovered and ultrafiltered through a 500 KDa hollow fiber filtration system and a 250 mL reverse osmosis (RO) wash and then reduced to ~2 mL to concentrate the viruses in the sample.

10.5.1.3   Preparation for IVDS

The solution was prepared for IVDS analysis by a 1:10 dilution with ammonium acetate (AA) as the salt for controlled conductivity. Each sample was filtered through W-41 20 μm paper or a 0.45 μm membrane filter to remove fatty and pollen residues that tend to float in a solution after centrifugation. The IVDS saved a five-scan average in its database. Charts and tables were created from data exported by the IVDS.

Virus detection by the IVDS is based on nanoparticle movement through a charged atmosphere. The particles display mass:charge ratios that can be converted easily to size data. The movement through the charged atmosphere allows the sorting of the particles in the sample by size, after which the condensed particles counter (CPC) unit counts particles that meet its calibrated size standard.

10.5.2   OBJECTIVES

10.5.2.1   Essential Oils as Viral and Nosema Controls in Honeybees

This study was intended to establish whether essential oils are effective as virus management tools in honey bees. We planned to evaluate the individual oils listed in Table 10.1 and the total combination of essential oils to determine if whether any mixture controlled viruses more effectively of the synergy of the oils together was better for controlling viral loads. Under a confidential disclosure arrangement, the treatment colonies were to be compared to control colonies and to each other to evaluate the efficacies of the oils.

10.5.2.2   Viral Load Relationship to Pests, Parasites, Diseases, and Stress of Managed Honeybees

This analysis was a larger effort, but the same techniques were to be used to evaluate the variables and develop dose response or cause-and-effect curves. We hoped that the viral relationships to the factors investigated would provide a new perspective for research in this area.

10.5.3 Challenges

This project presented many variables in techniques; some were sufficiently feasible to become standardized. Their utility was confirmed by others and used for other purposes. We simply shipped live bees in plastic bags and sent them in USPS flat-rate boxes. This method was inexpensive and easy for beekeepers and the samples arrived at the laboratory safely.

We had to resolve time-of-day collection issues before we could prepare samples. We risked losing or misrepresenting data because of diurnal variations or using different aged bees in our analysis. The problem was that younger bees do not carry the mature viral loads that can be seen in older bees. We addressed this with consistency and found no significant differences in viral loads based on the time of day of collection.

Despite these considerations, the first set of test and control bees all died without application of essential oils and no samples were sent to BVS for analysis. The second set of tests and controls failed because a beekeeper backed out of the experiment and caused a 4-month delay in starting the experiment. Perseverance, however, prevailed and 10 colonies were processed over several months along with a control set for a total sample set of 80. We discovered that these early colonies were already at high-intensity virus and Nosema levels. The results appeared inconclusive and seemed to indicate that essential oils could not prevent or cure Nosema.

The data indicated, however, that essential oils may reduce viral loads even starting at high intensity levels. The project continued with additional funding from the California State Beekeepers Association and Project Apis m. This support provided additional control and low viral load specimens and allowed us to evaluate low Nosema levels.

10.5.4 Virus Relationship to Stress

By using the interrelationships of honeybee health, stress factors, immune responses, and latent and active viral loads, we discovered a relationship between virus intensity and honeybee stress quantification.

By using the IVDS to detect and count viruses, we discovered that viral diversity and intensity in honeybees changes with increasing or decreasing stress levels. The quantification and tracking of viral loads continued over time. We began to see viral load as a function of stress placed on honeybees. The quantification of stress covered all phases of beekeeping from migratory practices, nutrition, pesticide exposure, chemical treatment, weather, and mono-crop pollination. Stress management is needed in the scope of commercial beekeeping but stress is difficult to quantify. The data we recorded and presented indicate a direct relationship of viral load to stress exposure. Some bees carry higher viral loads than others.

The results of our study led to a recommendation to growers and beekeepers to monitor bee viral loads at regular intervals as a measurement of bee health. We tied together the monitoring of viral loads as a means to quantify stress from sources such as shipping, migratory practices, pests, fungi, viruses, nutrition, water, over-population, and weather to name a few.

10.5.5   RESULTS AND DISCUSSION

- The data originated from nine hives.
- Collections were made three times daily (morning, afternoon, and evening).
- A total of 27 samples was taken for each size class.
- The intent was to determine patterns associated with time of day in the counts.
- Convincing evidence of the Sacbrood virus (SBV) was found in every hive.
- Evidence of numbers of virus particles present at different times of day was unconvincing.
- Counts for each hive were plotted against size and for each time of day (Figure 10.2).
- The peak associated with the Sacbrood virus is apparent in each panel.
- Clear differences in counts based on time of day are not visible.
- Morning counts are not greater than afternoon and evening counts.

BVS processed over 3,500 samples from beekeepers around the nation. The volume of data and the correlation of viral loads to seasonal changes, transportation overwintering, feeding, and the use of essential oils indicates the need for conclusive evidence to support practical application of the methods, management, and treatments used. BVS gathered observational data suggesting that essential oils are effective in the reduction of viral loads.

The data reveal that bee colonies with increasing viral loads respond favorably when treated with combinations of essential oils. The viral loads dropped and remained low for approximately 60 days. The weaknesses of the data arise from the lack of experimental methods and controls that provide reliable proof and details of application that would allow beekeepers to be confident about the outcomes of treatments.

The effort to bring scientific rigor to this project proved difficult. Even so, the results are promising and treatment combined with monitoring may produce additional ideas and a solution. Samples treated with essential oils generally started with higher viral counts than untreated samples. These higher counts decreased following treatment. Samples collected in August had higher viral loads for both treated and untreated samples when compared to samples collected in September. Both treated and untreated samples maintained high Nosema counts that did not change of the course of the study.

All samples were from the same apiary. Other factors should be considered when making a study of this nature. Remember that virus loading can be characterized by the use of IVDS, that is the constant. Controlling all the variables is the challenge for these types of studies. It was not clear for example, if the honeybees were treated prior with essential oils or indeed if all the honeybees were treated prior to testing.

Treated samples show overall higher mean counts for DWV, KBV, and IAPV with a near equal count for SBV. The non-treated group show higher counts for ABPV and CPV.

10.6 SNAPSHOTS OF VIRAL LOADS IN HONEYBEES

10.6.1 Objectives

One of the first projects involving honeybees was obtaining snapshots of viral loading in a variety of situations to develop baseline data for comparisons with subsequent measurements. The following observations were made:

- Viral loads of honeybees from colonies from coast to coast across the United States were monitored.
- Impacts of various treatments of honeybee colonies on viral loads over time were monitored.
- Initial data suggested that treatment with essential oils may have produced some beneficial impacts on viral loads.
- Untreated colonies exhibited strong frame strength while viral loads increased; this may indicate other influences not tested here but worth monitoring.
- Changes in viral loads are related to stress factors affecting bee health.

Bees have shown that they can maintain their viruses at a latent or manageable state without showing any symptoms of a specific virus. Stress is a problem in maintaining their viral loads. The more stress honeybees encounter, the harder it is for them to maintain a normal state and eventually they will show symptoms of viruses.

One early observation has been that virion counts are generally very low in young bees. They do not seem to have mature viral loads and thus may not represent the status of a colony based on normal levels of counts and diversity. In colonies that are declining in strength and health, we see younger bees develop as the older ones die or disappear as in the symptoms of colony collapse disorder (CCD).

BVS requested sampling of the oldest bees or foragers first in an attempt to standardize the data sets. Specific samples may differ, but we found that granting this request was not always possible. Samples were processed as delivered with appropriate notes indicating the specific circumstances. All samples involved multiple variables that could have exerted effects on the honeybees and their viruses.

10.6.2 Results

We found that the single bee and the feral bees demonstrated distinct viral detections. An unlabeled vial A yielded a distinct peak at 32.2 nm (Sacbrood virus). The single bee distinctly indicated Kashmir bee virus (KBV) at 23.3 nm. Israeli acute paralysis virus (IAPV) was indicated at 25.9 nm and a peak at 28.9 nm was not associated to a specific virus.

The counts seemed to be generally high or low as shown in detail in the charts. The lack of distinctive peaks in the samples certainly stood out. The individual peaks were marked and recorded for reference. The bees showing peaks representing viruses appear to be unstressed or the viruses are in a latent state. The bees appear to be doing well based on analysis of viral diversity, intensity, and the non-distinctive nature of the peaks.

Continued tracking over time will show the changes in bee stress as demonstrated by the viral load graphs and charts. The viral counts are typical and none of the peaks offers any indication of an unusual viral load with the exception of the single bee and feral bee samples. We maintained a running history of the many particles and viruses detected. Individual samples could be compared with this history to determine a situation within a temporal reference. Figure 10.1 demonstrates this capability. A sample is superimposed upon a temporal record of the average virus loading.

A 1.7-g sample showed peaks indicated an elevated count of deformed wing virus (DWV) at 20.2 nm but no distinct peak that might indicate an outside stress such as honey flow. The virus remains under control, perhaps by the bee immune system. This is the same observation made for IAPV at 25.0 nm and Sacbrood virus (SBV) 31.1 nm. This sample is indicative of young bees without mature viral loads although there is evidence of these viruses in the colony.

The peaks from a 1.1g sample revealed IAPV at 25.0 nm, an undesignated virus at 28.9 nm, and chronic paralysis virus (CPV) at 35.9 nm. The IAPV is more concentrated and consistent over all five scans. The counts are midlevel to historical levels and appear to indicate that the samples are from older bees with mature viral loads that are managed in a healthy colony.

A 1.1g sample showed detections of DWV (21.7 nm) and CPV (37.2 nm) that were consistent in all five scans. The counts are in the medium range and not significantly distinct in the chart. The sample appeared to contain older bees with mature viral loads that were managed well in a healthy colony.

A 6.0g sample noted detections of DWV (21.7 nm), KBV (23.3 nm), acute bee paralysis virus (ABPV, 26.9 nm), and (CPV, 35.9 nm). The peaks are present and consistent in all five scans. The counts are midlevel and do not demonstrate elevated stress in the colony. The KBV and DWV levels are elevated but do not show a corresponding increase in the salt curve. This colony sample appears to consist of younger bees with elevated viral loads. The colony is healthy but may be showing signs of stress.

Another 6.0g sample showed distinct and consistent peaks at DWV (20.2 nm) and KBV (23.3 nm) over all five scans. The counts were in the very low range and not significantly distinct. The sample appeared to come from younger bees without mature viral loads although their loads could be measured. Please read the notes on young bees in Section 10.6.1.

A small weight sample (<0.1g) from a single bee was examined. Peaks were noted for KBV (23.3 nm), IAPV (25.9 nm), and an unassociated virus (28.9 nm). These counts were consistent and in very low but distinct concentrations. The presence and counts of these viruses in a single bee tend to indicate high stress level and an unmanaged viral infection. It is unclear which of these viruses had more impact on the health of this bee, but their presence clearly indicates an elevated load.

A 0.4-g sample showed very low (into background) virus levels. The single consistent detection was of KBV (22.5 nm). This sample would appear to be from a young bee with no significant viral load.

10.6.3   PROJECT SUMMARY

This project provided snapshots of the viral loads in honeybees. The first observation was generally low viral loading among many samples. Some of these samples may represent historical snapshots for future comparisons because the perils for honeybees continue without a suitable solution for the foreseeable future.

It would be very beneficial to map the distribution, frequency, diversity, and intensity of the viral loads in honeybee regions to determine trends and relationships of viral loads to other factors; such information could also serve as tools for management of colonies.

It is well documented that bees carry viruses. When bees are stressed by various conditions (shipping, weather, poor nutrition, pests, Nosema, pesticides, and pesticide treatments, their viral loads (diversities and counts) change. Latent viruses often become active and otherwise normal counts increase, perhaps enough to prevent the bees from thriving and eventually kill them.

10.7   CONCLUSIONS

A group of charts best summarizes the use of IVDS with honeybees by depicting dose response curves for treating Nosema (Figures 10.8 through 10.10). Figure 10.9 clearly shows that the initial impact upon application of a fungicide is a rapid increase in virus loading. The virus loading then decreases below the first measurement and indicates a return to healthy status. The treatments can be varied by the fungicide chosen, the dose used, and other factors such as feeding practices.

The cause-and-effect curve can be expanded to demonstrate the effects of temperature and relative humidity. The effects of types, durations, and amounts of feeding can also be explored. All these cause-and-effect curves can be used to determine impacts on honeybee health and exploring measures to promote honeybee health.

Figures 10.1 and 10.3 show the advantages of measuring and monitoring virus loading over time. It is possible to correlate these temporal charts with local environmental factors such as temperature and relative humidity. The peak detection frequency (both annual and seasonal) allows the prediction of natural changes that impact viral loading and can be used to plan honeybee maintenance activities to minimize the effects.

Similarly, the ability to monitor a single colony (Figures 10.4 to 10.7) allows close monitoring of valuable hives and testing of management practices on a single colony before applying them to an entire yard. This capability enables hives on the edge of a yard to be compared to hives throughout the entire yard. Changes in viral loading can then be used to deal with an invasion by an unknown agent such as weather or insecticide use. Consistent monitoring would permit outside influences to be noticed immediately and handled by proper action.

The measurement of the relative abundance of viruses in honeybees is another important factor. Knowing what the honeybees can tolerate naturally allows workers to focus on other issues that may be more important. Virus biomarkers can be used to make appropriate management decisions and promote healthy honeybees.

11 Pulsed Lamp Decontamination

11.1 INTRODUCTION

The ability to decontaminate microorganisms with a broadband pulsed light system has been demonstrated (Wick et al. 1998b; Wick et al. 2012c). This chapter will review the effects of the pulsed lamp decontamination on MS2 bacteriophage and tomato bushy stunt virus (TBSV). IVDS allows the data to be analyzed quickly, without the use of historical microbiological techniques. When such microbiological processes were used on *Escherichia coli* bacteria, mounds of Petri dishes were required and the results were not known until the cultures grew overnight. When other more fastidious microbes were cultured, the analysis time was usually even longer.

A pulsed light source is characterized as a rapid decontamination device. Although it usually serves as a broadband white light source, we found it desirable to evaluate selected wavelengths by filtering them and allowing some to pass through various optical filters. The effect of this treatment was measured by recording the counts of MS2 and TBSV (Wick 2010b).

Exposure to filtered and unfiltered pulsed lamp exposure was provided by a full spectra pulse lamp. Various filters were placed in the path of the lamp to determine effects of wavelength on the decontamination of MS2 and TBSV. Sample virus content was analyzed before and after exposure to the pulsed light.

Another purpose of this experiment was to demonstrate the feasibility of using the IVDS to collect data about these viruses and develop a dose response curve, much like those developed for temperature and pH. The rapid analysis capability of the IVDS was expected to allow rapid adjustments of the method to enhance the effectiveness of the treatment. Likewise, the immediate response provided for a quick assessment of the effectiveness of the various wavelengths and enabled the correct wavelength required for decontamination to be selected.

Earlier work with bacteria (Wick 1998b) demonstrated the effectiveness of broadband pulsed light to decontaminate microbes. The light was found to be very effective and capable of sterilizing a Petri dish containing more than 10^8 microbes in a single pulse. Using this method to examine viruses was the next step.

11.2 METHODS

11.2.1 PULSED LIGHT

The pulsed lamp is housed in a stainless steel enclosure (24 in. × 18 in. × 20 in.) centered on the inner top panel parallel to the opening. The enclosure is interlocked to disarm the control panel in the event the door is open or opens during experimentation. The electronics for activating the lamp are housed adjacent to the enclosure. The control panel allows the repetition rate to be set from 1 to 9 pulses. More pulses can be activated by reactivating the start button (Wick 2012).

Samples were centered under the lamp in the enclosure. A filter hood was used in all experiments to allow the placement of various filters in the light path to examine the effects of different wavelengths. The filter hood schematic and distance from the lamp are shown in Figure 11.1. The virus samples were exposed in a six-well well microscope slide (Figure 11.2) measuring 2 in. × 3 in. × 0.25 in. thick. Each well is 3/64 in. deep.

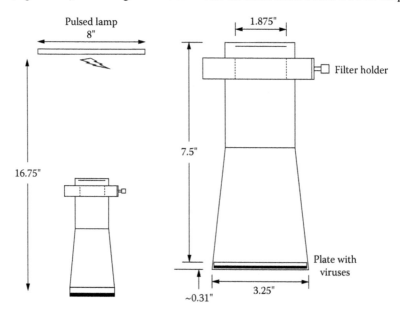

FIGURE 11.1   Pulsed lamp setup with filter hood.

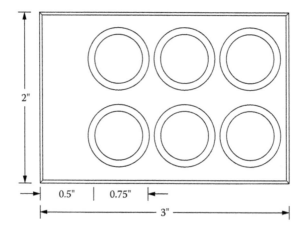

FIGURE 11.2   Sample holder for pulsed light exposure.

11.2.2 Filter Data

Various filters were used to attenuate the energy or wavelength from the pulsed lamp before impingement on the virus samples. The filters were measured by the US Army Center for Health Promotion and Preventive Medicine (USACHPPM) to determine actual wavelength cutoffs.

Large and small Petri dishes, the areas of the Petri dishes with logos, and the well sample holders were also measured. The Petri dishes allow transmittance in the visible spectrum, while blocking ultraviolet (UV) light below 310 to 325 nm at 50% transmittance (T). The Petri dishes should be uncovered when used in experiments with pulsed light effects. The well slide blocks UV light below 325 nm at 50% T, and no large effects were expected from secondary bounces of light from the stainless surface below the slide. The pulsed lamp was measured with two radiometers (IL 1400 and IL 1700) with various bandpass filters. The measured fluence measurements with various filters are shown in Tables 11.1 and 11.2.

TABLE 11.1
IL 1400 Radiometer Results (Set B, SEL 623, S/N 647)

Filter	Orientation	Pulses	Fluence (mJ/cm²)
NBP 400.2 nm		4	1.4
NBP 632.8 nm	Mirror up	3	1.9
NBP 632.8 nm	Mirror up	3	2.0

TABLE 11.2
IL 1700 Radiometer Results (Set A, SED 623 S/N 180)

Filter	Orientation	Pulses	Fluence (mJ/cm²)
NBP 400.2 nm		3	1.45
NBP 400.2 nm		3	1.45
NBP 400.2 nm	Filter reversed	3	1.49
NBP 254 nm	Mirror up	3	0.71
NBP 254 nm	Mirror up	3	0.66
NBP 254 nm	Mirror down	3	0.77
NBP 632.8 nm	Mirror up	3	1.95
NBP 632.8 nm	Mirror down	3	1.73
NBP 632.8 nm	Mirror down	3	1.7
NBP 632.8 nm	Mirror up	3	1.5
NBP 632.8 nm	Mirror up	3	1.82
NBP 632.8 nm	Mirror down	3	1.98

11.2.3 IVDS

The IVDS used was the same device described in Chapters 1 and 2. Direct counting of samples was possible because they were prefiltered and prepared for the counting unit before the experiment. The samples were inserted into the pulsed light apparatus, exposed, and then directly analyzed by IVDS. The whole process took minutes.

11.2.4 EXPERIMENTAL ARRANGEMENT

MS2 bacteriophage #04, stock sample, was measured with the IVDS prior to each set of pulsed light exposures and each time there was a new dilution from the stock (Figure 11.3). The wells of the microscope slide were filled with 100 μL of solution. For multiple exposures, the samples were left in the wells until the target number of pulses was reached. For example, for one sample with 10 pulses and another with 20 pulses, two wells would be filled. The two wells would then be pulsed 10 times and one well emptied after pulsing. The remaining well would then be subjected to another 10 pulses, and the sample would be removed. The second well would then have received 20 pulses. Multiple well samples could then be compared as the same stock sample was exposed to differing pulse counts.

These general procedures were followed for each test and modified as noted below. An allotment of MS2 bacteriophage #04 was used to fill each well in the sample slide with 100 μL MS2 or tomato bushy stunt virus (TBSV) as appropriate. The slide was then located under the filter holder with or without a filter, depending on the test. Each slide was pulsed with 5 pulses and one well was removed after each pulse session, resulting in samples undergoing 5, 10, 15, 20, 25, and 30 pulses each.

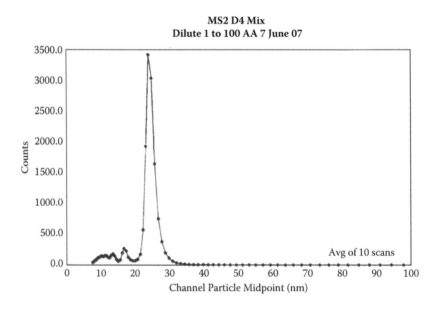

FIGURE 11.3   MS2 #04 stock dilution sample.

The samples were analyzed with the IVDS, and the results were depicted in figures and tables as described below.

11.3   RESULTS

The results of subjecting MS2 bacteriophage and TBSV samples to pulsed light under various filtering conditions are described below.

11.3.1   MS2 Bacteriophage

Results for filtered and unfiltered samples are detailed in this section.

MS2 Bacteriophage (Unfiltered)—To determine the effects of the pulsed light on a sample of MS2 bacteriophage, the slide was placed after filling under the filter holder (no filter). Each slide was subjected to 5 pulses and one well was removed after each pulse session, resulting in samples with 5, 10, 15, 20, 25, and 30 pulses each. The samples were analyzed with the IVDS, and the results are shown in Figure 11.4. The peak counts are shown in Table 11.3.

MS2 with GG400 Filter—Three wells in a well slide were filled with MS2 #04 after it was diluted from the stock sample. The slide was placed under the filter holder with a GG400 filter in place. The samples were pulsed for 10, 20, and 30 pulses each. The IVDS results of the pulsed samples are shown in Figure 11.5. The top section of the IVDS scan is shown for clarity of defining the peak counts. The peak counts are shown in Table 11.4.

MS2 with WG320 Filter—Three wells in a well slide were filled with MS2 #04 after its dilution from the stock sample. The slide was placed under the filter holder

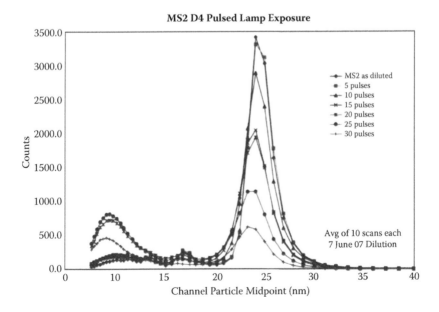

FIGURE 11.4   MS2 pulsed sample (no filter).

TABLE 11.3
Pulses and Peak Counts for MS2 #04 (Unfiltered)

Pulses	Peak Count	Reduction (%)
0.0	3420	—
5.0	3317	3
10.0	2892	15
15.0	2056	40
20.0	1951	43
25.0	1156	66
30.0	635	81

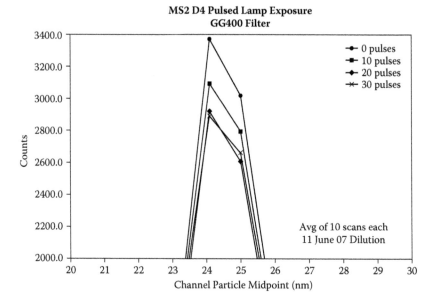

FIGURE 11.5   MS2 #04 pulsed sample (GG400 filter).

with the WG320 filter in place. The samples were pulsed for 10, 20, and 30 pulses each. The IVDS results of the pulsed samples are shown in Figure 11.6. The top section of the IVDS scan is shown for clarity of defining the peak counts. The peak counts from the IVDS graph for the WG320 filter are shown in Table 11.5.

MS2 with WG225 Filter—Three wells of a well slide were filled with MS2 #04 after its dilution from the stock sample. The slide was placed under the filter holder with the WG225 filter in place. The samples were pulsed for 10, 20, and 30 pulses each. The results are shown in Figure 11.7. The peak counts from the IVDS graph for the WG 225 filter are shown in Table 11.6.

MS2 with UV Cold Mirror Filter—Three wells of a well slide were filled with MS2 #04 after its dilution from the stock sample. The slide was placed under the

TABLE 11.4
Pulses and Peak Counts for MS2 #04
with GG400 Filter

Pulses	Peak Count	Reduction (%)
0.0	3370	—
10.0	3100	8
20.0	2927	13
30.0	2892	14

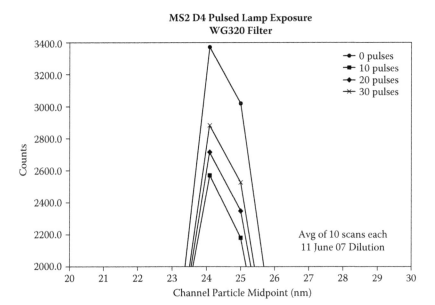

FIGURE 11.6   MS2 #04 pulsed samples (WG320 filter).

TABLE 11.5
Pulses and Peak Counts for MS2 #044
with WG320 Filter

Pulses	Peak Count	Reduction (%)
0.0	3370	—
10.0	2574	24
20.0	2720	19
30.0	2871	15

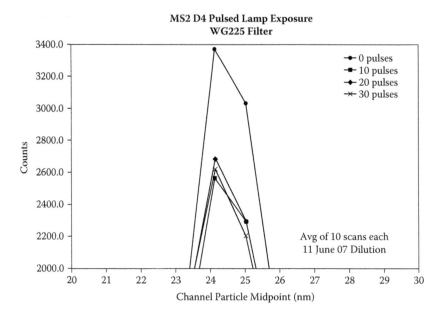

FIGURE 11.7   MS2 #04 pulsed samples (WG225 filter).

TABLE 11.6
Pulses and Peak Counts for MS2 #04
with WG225 Filter

Pulses	Peak Count	Reduction (%)
0.0	3370	—
10.0	2549	24
20.0	2674	21
30.0	2607	23

---

filter holder with the UV cold mirror filter in place. The samples were subjected to 10, 20, and 30 pulses each. The results are shown in Figure 11.8 and peak counts are listed in Table 11.7.

MS2 with Infrared (IR) Suppression Filter—Three wells of a well slide were filled with MS2 #04 after its dilution from the stock sample. The slide was placed under the filter holder with the IR suppression filter in place. The samples were subjected to 10, 20, and 30 pulses each. The analysis is shown in Figure 11.9 and peak counts are shown in Table 11.8.

11.3.2   TBSV

Results for filtered and unfiltered samples are detailed in this section.

TBSV (Unfiltered)—Three wells in the well slide were filled with TBSV after its dilution from the stock sample. The dilution is shown in Figure 11.10. The slide was

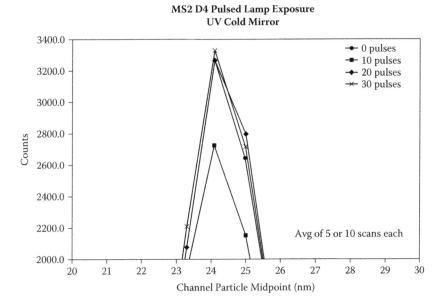

FIGURE 11.8   MS2 #04 pulsed samples (ultraviolet cold mirror).

TABLE 11.7

Pulses and Peak Counts for MS2 #04 with UV Cold Mirror Filter

Pulses	Peak Count	Reduction (%)
0.0	3269	—
10.0	2726	17
20.0	3279	+0.3
30.0	3334	+2

placed under the filter holder with no filter in place. The samples were pulsed for 10, 20, and 30 pulses. The IVDS analysis of the pulsed sample is shown in Figure 1.11. The peak counts for the unfiltered sample are shown in Table 11.9.

TBSV with WG225 Filter—Three wells in a well slide were filled with TBSV after its dilution from the stock sample. The slide was placed under the filter holder with a WG225 filter in place. The samples were pulsed for 10, 20, and 30 pulses each. The IVDS analysis is shown in Figure 11.12. The peak counts for the WG225 filter are shown in Table 11.10.

11.4   DISCUSSION AND CONCLUSIONS

The pulsed light exposure reduced the IVDS counts of MS2 and TBSV. More light applied to samples produced greater reductions in counts. The WG225 filter allowed

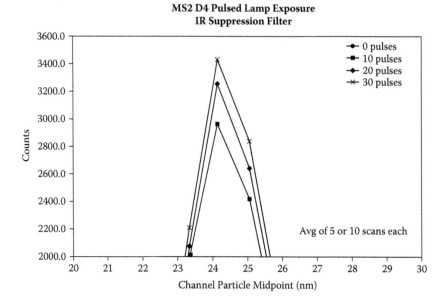

FIGURE 11.9   MS2 #04 pulsed samples (infrared suppression).

TABLE 11.8
Pulses and Peak Counts for MS2 #04
with IR Suppression Filter

Pulses	Peak Count	Reduction (%)
0.0	3269	—
10.0	2974	9
20.0	2978	9
30.0	3443	+5

the most UV light to pass and produced the largest reduction in counts except for the unfiltered pulsed light exposure.

The reduction in counts for 30 pulses did not always result in linear reductions in counts. One explanation may be that the well plate used for the virus solutions is not symmetrical; the wells are shifted off center.

The large pulse counts, typically 30 pulses, were always in the wells nearest the edge of the slide. The position of the well edge against the edge of the filter holder may have produced a shadowing effect from the holder that reduced the pulsed lamp exposure. This may explain why the non-linearity of the higher pulse exposures did not demonstrate increasing reductions of counts in some cases.

Further experimentation is warranted to determine whether the shadowing affected the results. Future experiments should also utilize symmetrical well plate for exposing samples to the pulsed light. Other variables should be considered,

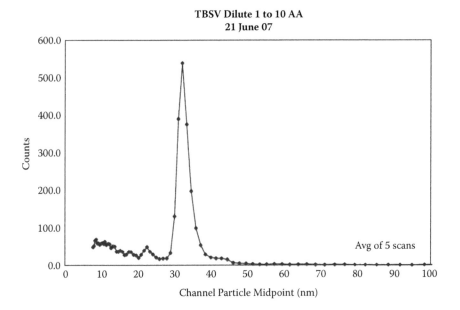

FIGURE 11.10   TBSV stock sample.

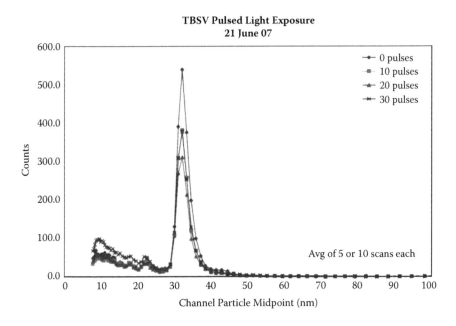

FIGURE 11.11   TBSV stock and pulsed samples (no filters).

TABLE 11.9
Pulses and Peak Counts for TBSV
(Unfiltered)

Pulses	Peak Count	Reduction (%)
0	539	—
10	381	29
20	311	42
30	378	30

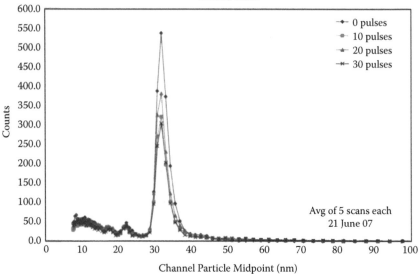

FIGURE 11.12   TBSV stock and pulsed samples (WG225 filter).

TABLE 11.10
Pulses and Peak Counts for TBSV
with WG225 Filter

Pulses	Peak Count	Reduction (%)
0	539	—
10	324	40
20	384	29
30	306	43

for example, the shadowing of the viruses in solution from each other, that is, the decontaminated viruses on the contact side of the exposure may have attenuated further exposure.

The pulsed light exposure reduced the counts of MS2 and TBSV. IVDS analysis demonstrated that the higher the number of pulses from the light, the larger the reduction in counts in a sample. Greater reductions in counts were also noted when the filters allowed more UV light to penetrate to samples. This does not indicate that reduced counts are attributed only to UV wavelengths. Further experimentation may resolve this question. Results indicate that the killing effect observed from the pulses is additive—hardy microbes may be killed by increasing either the power of the lamp or the number of pulses.

12 Current and Future Trends

12.1 INTRODUCTION

The potential applications of IVDS in the current configuration are numerous; likewise the applications of the next generation IVDS and RapidDx are even more numerous. The future is open to those who have imaginations and can dream of having a near real-time virus detector that is not based on molecular biology or complicated chemistry, but rather is a physical-based approach. As I noted earlier, the IVDS can sit on a shelf with minimum attention and be turned on and operated quickly when needed; that means no reagent shelf lives to worry about.

12.2 POTENTIAL APPLICATIONS OF IVDS

The IVDS has several potential applications in the medical and defense arenas including homeland defense. In the medical arena, it can be valuable in the clinical and research areas. As a research tool, it can be used to detect the presences of unknown viruses. For example, after the start of the AIDS epidemic, many years of research were required to determine that the culprit was an unknown (at the time) virus. It is very possible that the AIDS virus could have been found by the IVDS if the system existed at the time, and research for controlling the disease could have started much earlier.

Another clinical use for the IVDS comes to mind. HIV-positive patients who control their disease by medication must undergo blood tests every few months; the concentration of the virus must be tested to ensure that their medication regimes are still effective. This test is fairly expensive and labor intensive. The IVDS could simplify the testing and bring the cost down.

In addition, the HIV mutates periodically and a patient's medication regime must be changed. It is probable that a mutated virus is of a different size and/or generates a different breakdown product. The IVDS could possibly detect the mutated virus early and allow a physician to start a new medication regime before the new virus caused too much patient damage. Early detection of hepatitis is another potential clinical application for the IVDS on the basis that the five types of hepatitis viruses are very different in size.

Viruses can potentially be used as biological weapons. At present, we do not have any rapid virus detection capabilities. Mating an appropriate aerosol collector with the IVDS can fill this gap in our detection capabilities. In the medical field and the defense arena, we need to characterize the known viruses. Concerted effort is required to characterize the sizes and breakdown products of the known viruses and develop a database that will contain all relevant information.

Similarly, we need to characterize the background atmosphere in terms of the viruses and virus-like particles it contains naturally to determine how this natural background varies through time and space. IVDS can be a first step in this activity. After the samples have been collected and characterized by IVDS, it is simple to complete analysis and identify and classify the particles by other means, for example by mass spectrometry proteomics. This technique has a wide capability to identify all three classes of microbes: viruses, bacteria, and fungi (Wick 2014).

12.3   NEW RAPIDDX INSTRUMENT

The development of the RapidDx-IVDS is exciting from many aspects. First, it upgrades the current model by improving resolution and automation. The addition of automation is welcome because it allows unattended sampling and analysis. The throughput of this new instrument will be improved greatly. Furthermore, the upgrade of the condensation particles counter (CPC) to a CCD device for counting viruses will reduce sampling time by several minutes per sample. The unit is also rugged and portable.

Improved software will facilitate the distribution of results from the RapidDx instrument. One desirable feature is the system's ability to automatically sample, analyze, and record results. The results are recorded in computer files and can be distributed via electronic means. This means that multiple units working in an area can coordinate their results and send a summary report to one or more central locations where they can be monitored. An alert (e.g., change in normal particle loading) would be noted by the software and alarms can be sent to the relevant parties for quick and responsive management.

12.3.1   CONCEPT OF OPERATION AND APPLICATIONS

The concept of operation depends on a number of applications as detailed below. A critical component would be the development of more virus detection techniques by life science laboratories through research and development deployments of DMA prototype systems in medical research and industrial quality control settings, for example, applications in process diagnostics for synthetic vaccine synthesis. Commercial deployments in these areas would mean highly rugged automated systems operable by minimally trained technicians. It would no longer be possible for a vaccine to contain hidden unrelated viruses; vaccines could be checked for purity easily. Urgent applications for IVDS technology include the following:

- Point-of-care medical diagnostics. Single-stage differential mobility analyzer (DMA) deployment for virus screening can revolutionize life sciences. Diagnostic applications would be utilized by military and civilian hospitals and in clinical and pharmacy settings following Food and Drug Administration approval.

- Transportation security screening portals for passengers, luggage, and vehicle traffic. The supersensitive high-resolution system could be used to monitor airborne vapors directly without need for swabbing to transfer particulate samples.
- Ready reserves. Specialized DMA equipment intended for rapid deployment to first responders to address epidemics and other public health catastrophes could be stored in stand-by mode in a clean, dry warehouse facility, ready to go when needed. Set-up could be achieved in minutes with no need to prepare and deliver reagents or other supplies.

12.3.2   ELECTRONIC AMPLIFICATION

Conventional polymerase chain reaction (PCR) methods of microbe identification rely on chemical amplification for one-at-time genetic detections of species and strains. The IVDS, especially the new RapidDx combination of electrospray (ES), DMA, and CCD with solid-state electronics, employs generic (non-chemical) electronic amplification to detect all virus species simultaneously and generate true negative or positive results. Furthermore, an ion current signal detection system (ES, DMA, CCD) is inherently easier to integrate with computer electronic and data communications components.

12.3.3   MINIATURIZATION

There is considerable scope for further miniaturization of the ES-DMA-CCD systems. Designing for manufacturability could aid this effort.

12.3.4   BALANCE OF SYSTEM

A major advantage is that the balance of system engineering for hardware, electronics, and software can be shared across multiple applications of the platform, resulting in commonality of parts and spares to minimize support and maintenance requirements. Multiplexed DMA subsystem deployments such as the proposed tandem DMA could share key system components and subassemblies such as power supplies and control systems. Common systems components can also support future miniaturized versions of this equipment.

12.4   THE FUTURE

The future for the IVDS instrument and techniques such as mass spectrometry proteomics is fantastic when you consider the rapid advances in electronics and software in the past 10 or so years. It is easy to envision these devices used widely for medical diagnostics and general surveillance applications. The devices will be completely portable and make rapid and individual diagnostic and other techniques to be widely available.

A first step in this direction may be the automation of these instruments and placing them in a central location such as a drugstore, where customers could perform their own tests for microbes. This would make secondary treatment easier because each individual would know what microbe he or she carried.

In the event of a pandemic, healthcare facilities would not be so taxed if individuals knew their own viral status. At present, a pandemic would swamp healthcare facilities simply because many individuals will have common symptoms and seek care. Individuals who had no symptoms and knew their microbial status was not harmful would not have to seek help at overloaded healthcare facilities. Furthermore, quick assessments of microbial status at healthcare facilities quickly treat individuals who need help.

We must consider the overuse of antibiotics. In recent years, we have seen the evolution of antibiotic-resistant microbes. The reasons that microbes develop antibiotic resistance is not relevant to this discussion but the solution to the problem is. The IVDS is a simple answer to screening people for viruses. Physicians generally do not prescribe antibiotics for viruses. IVDS screening would identify people with viruses and they would not be given antibiotics that would not help them. Current practice is to administer general antibiotics and hope the symptoms improve; improvement may or may not be a function of the treatment. The ability to screen virus from non-virus infections would be a big step forward.

The next step is to identify the microbes using a robust and broad-spectrum device to give an accurate diagnosis by identifying the relevant microbe. Exact identification of bacteria to strain where possible would allow precise targeting of antibiotics and thus ensure effectiveness of treatment. This capability would also help alleviate the overprescription and misuse of antibiotics. Of course, bacterial identification involves many variables beyond knowing the microbe responsible for an infection. Someday, advances in technology will allow us to buy and use complete diagnostic systems in supermarkets as we pass through checkout lines!

References

Baron, C., N. Domke, R.M. Beinhofe et al. 2001. Elevated temperature differentially affects virulence, VirB protein accumulation, and T-pilus formation in different *Agrobacterium tumefaciens* and *Agrobacterium vitis* strains. *Journal of Bacteriology 183*: 6852–6861.

Bothner, B., A. Schneemann, D. Marshall et al. 1999. Crystallographically identical virus capsids display different properties in solution. *Nature Structural & Molecular Biology 6*: 114–116.

Bouquet, H. 1763. Bouquet Letters, MSS 21634:295 and 231, British Library, London.

Bromenshenk, J.J, C.B. Henderson, C.H. Wick et al. 2010. Iridovirus and microsporidian linked to honeybee colony decline. *PLoS ONE 5*: doi:10.1371/journal.pone.0013181

Cargile, B.J., S.A. McLuckey, and J.L. Stephenson. 2001. Identification of bacteriophage MS2 coat protein from *E. coli* lysates via ion trap collisional activation of intact protein ions. *Analytical Chemistry 73*: 1277–1285.

Dai, Y., L. Li, D.C. Roser et al. 1999. Detection and identification of low-mass peptides and proteins from solvent suspensions of *Escherichia coli* by high performance liquid chromatography fractionation and matrixassisted laser desorption ionization mass spectrometry. *Rapid Communications in Mass Spectrometry 13*: 73–78.

Fader, R.C., L.K. Duffy, C.P. Davis et al. 1982. Purification and chemical characterization of type I pili isolated from *Klebsiella pneumoniae. Journal of Biological Chemistry 25*: 3301–3305.

Fleischmann, R., M. Adams, O. White et al. 1995. Whole-genome random sequencing and assembly of *Haemophilus influenza* Rd. *Science 269*: 496–512.

Harris, S. 1992. Chemical and biological warfare. *Annals of New York Academy of Sciences 666*: 21–48.

Jensen, P.K., L. Pasa-Tolic, G.A. Anderson et al. 1999. Probing proteomes using capillary isoelectric focusing electrospray ionization–Fourier transform ion cyclotron resonance mass spectrometry. *Analytical Chemistry 71*: 2076–2084.

Karty, J.A., S. Lato, and J.P. Reilly. 1998. Detection of bacteriological sex factor in *E. coli* by matrix-assisted laser desorption–ionization time-of-flight mass spectrometry. *Rapid Communications in Mass Spectrometry 12*: 625–629.

Krishnamurthy, T., U. Rajamani, P.L. Ross et al. 2000. Mass spectral investigations on microorganisms. *Journal of Toxicology Toxin Reviews 19*: 95–117.

Kuzmanovic, D.A., I. Elashvili, C.H. Wick et al. 2003. Bacteriophage MS2: molecular weight and spatial distribution of protein and RNA components by small-angle neutron scattering and virus counting. *Structure 11*: 1339–1348.

Lei, Q.P., X. Cui, D.M.J. Kurtz et al. 1998. Electrospray mass spectrometry studies of non-heme iron-containing proteins. *Analytical Chemistry 70*: 1838–1846.

Morse, S.S., Ed. 1993. *Emerging Viruses*. New York: Oxford University Press.

Mouradian, S., J.W. Skogen, F.D. Dorman et al. 1997. DNA analysis using an electrospray scanning mobility particle sizer. *Analytical Chemistry 69*(5): 919–925.

St. Geme, J.W., J.S. Pinkne, G.P. Krasan et al. 1996. *Haemophilus influenzae* pili are composite structures assembled via the HifB chaperone. *Proceedings of the National Academy of Sciences of the USA 93*: 11913–11918.

Stimson, E., M. Virji, S. Barker et al. 1996. Discovery of a novel protein modification: alpha-glycerophosphate is substituent of meningococcal pilin. *Biochemical Journal 315*: 29–33.

Stockholm International Peace Research Institute. 1971. *The Problem of Chemical and Biological Warfare*, Vol. 1. New York: Humanities Press.

Temin, H.M. The high rate of retrovirus variation results in rapid evolution. In *Emerging Viruses*, S.S. Morse, Ed. New York: Oxford University Press, 1993, pp. 219–225.

Thomas, J.J, B. Bothner, J. Traina et al. 2004. Electrospray ion mobility spectrometry of intact viruses. *Spectroscopy 18:* 31–36.

Thomas, J.J., B. Falk, C. Fenselau et al. 1998. Viral characterization by direct analysis of capsid proteins. *Analytical Chemistry 70*: 3863–3867.

Tito, M.A., T. Kasper, V. Karin et al. 2000. Electrospray time-of-flight mass spectrometry of the intact MS2 virus capsid. *Journal of the American Chemical Society 122*: 3550–3551.

Tong, W., A. Link, J.K. Eng et al. 1999. Identification of proteins in complexes by solid-phase microextraction/multistep elution/capillary electrophoresis/tandem mass spectrometry. *Analytical Chemistry 71*: 2270–2278.

Ushinsky, S.C., H. Bussey, A.A. Ahmed et al. 1997. Histone HI in *Saccharomyces cerevisiae*. *Yeast 13*: 151–161.

Valegard, K., L. Liljas, K. Fridborg et al. 1990. Three-dimensional structure of the bacterial virus MS2. *Nature 345*: 36–44.

Webster, R.G. Influenza. In *Emerging Viruses*, S.S. Morse, Ed. New York: Oxford University Press, 1993, pp. 37–45.

Wick, C.H. 2002. Method and Apparatus for Counting Submicron-Sized Particles. US Patent 6,485,686 Bl.

Wick, C.H. 2002. Method and System for Detecting and Recording Submicron-Sized Particles. US Patent 6,491,872.

Wick, C.H. 2007. Method and System for Detecting and Recording Submicron-Sized Particles. US Patent 7,250,138 B2.

Wick, C.H. 2010. Detecting Bacteria by Direct Counting of Structural Protein Units or Pili by IVDS and Mass Spectrometry. US Patent 7,850,908 Bl.

Wick, C.H. 2011. Detecting Bacteria by Direct Counting of Structural Protein Units or Pili and Mass Spectrometry. US Patent 8,021,884.

Wick, C.H. 2012. Concentrator Device and Method of Concentrating a Liquid Sample. US Patent 8,146,446 Bl.

Wick, C.H. 2012. Virus and Particulate Separation from Solution. US Patent 8,309,029 Bl.

Wick, C.H. 2013. Virus and Particulate Separation from Solution. US Patent 8,524,155 B1.

Wick, C.H. 2013. Method and System for Sampling and Separating Submicron-Sized Particles Based on Density and/or Size to Detect Presence of a Particular Agent. US Patent 8,524,482 B1.

Wick, C.H., Ed. 2014. *Identifying Microbes by Mass Spectrometry Proteomics*. Boca Raton, FL: CRC Press.

Wick, C.H. and D.M. Anderson. 2000. System and Method for Detection Identification and Monitoring of Submicron-Sized Particles. US Patent 6,051,189.

Wick, C.H., D.M. Anderson, and P.E. McCubbin. 1999. *Characterization of the Integrated Virus Detection System (IVDS) Using MS2 Bacteriophage*. ECBC-TR-018 (AD-A364117). US Army Edgewood Chemical and Biological Center: Aberdeen Proving Ground, MD.

Wick, C.H., H.R. Carlon, R.L. Edmonds et al. 1997a. *Rapid Identification of Airborne Biological Particles by Flow Cytometry, Gas Chromatography, and Genetic Probes*. ERDEC-TR-443. US Army Edgewood Chemical and Biological Center: Aberdeen Proving Ground, MD.

Wick, C.H., H. Carlon, H. Yeh, and D. Anderson. 1998. *Quasi-Real-time Monitor for Airborne Viruses*. ERDEC-TR–45. US Army Edgewood Chemical and Biological Center: Aberdeen Proving Ground, MD.

Wick, C.H., M.D. Dunkel, R. Crumley et al. 1998. *Pulsed Light Device (PLD) for Deactivation of Biological Aerosols*. ERDEC-TR-456. US Army Edgewood Chemical and Biological Center: Aberdeen Proving Ground, MD.

Wick, C.H., R.L. Edmonds, and J. Blew. 1995. *Rapid Detection and Identification of Background Levels of Airborne Biological Particles.* ERDEC-TR-155. US Army Edgewood Chemical and Biological Center: Aberdeen Proving Ground, MD.

Wick, C. H, I. Elashvili, P.E. McCubbin et al. 2005. *Determination of MS2 Bacteriophage Stability at High Temperatures Using IVDS.* ECBC-TR-453. US Army Edgewood Chemical and Biological Center: Aberdeen Proving Ground, MD.

Wick, C.H., I. Elashvili, I., P.E. McCubbin et al. 2005. *Determination of MS2 Bacteriophage Stability at High pH Using IVDS.* ECBC-TR-472. US Army Edgewood Chemical and Biological Center: Aberdeen Proving Ground, MD.

Wick, C.H., I. Elashvili, R. Jabbour et al. 2006. Mass spectrometry and integrated virus detection system characterization of MS2 bacteriophage. *Toxicology Mechanisms and Methods 16*: 485–493.

Wick, C.H. and P.E. McCubbin. 1999. Characterization of purified MS2 bacteriophage by the physical counting methodology used in the integrated virus detection system (IVDS). *Toxicology Methods 9*: 245–252.

Wick, C.H. and P.E. McCubbin. 1999. Purification of MS2 bacteriophage from complex growth media and resulting analysis by the integrated virus detection system (IVDS). *Toxicology Methods 9*: 253–263.

Wick, C.H. and P.E. McCubbin. 1999. Passage of MS2 bacteriophage through various molecular weight filters. *Toxicology Methods 9*: 265–273.

Wick, C.H. and P.E. McCubbin. 2005. *Stability of IVDS Electrospray Module during Analysis of MS2 Bacteriophage.* ECBC-TR-462. US Army Edgewood Research, Development and Engineering Center: Aberdeen Proving Ground, MD.

Wick, C.H., P.E. McCubbin, and A. Birenzvige. 2006. *Determination of MS2 Bacteriophage Stability at High pH Using IVDS. ECBC-TR- 472.* US Army Edgewood Research, Development and Engineering Center: Aberdeen Proving Ground, MD.

Wick, C.H., P.E. McCubbin, and A. Birenzvige. 2006. *Determination of MS2 Bacteriophage Stability at Low pH Using IVDS. ECBC-TR-473.* US Army Edgewood Research, Development and Engineering Center: Aberdeen Proving Ground, MD.

Wick, C.H. and P.E. McCubbin. 2010. *Malvern Nano ZS Particle Size Comparison with the Integrated Virus Detection System (IVDS).* ECBC-TR-749. US Army Edgewood Research, Development and Engineering Center: Aberdeen Proving Ground, MD.

Wick, C.H. and P.E. McCubbin. 2010. *Capillary Diameter Variation for the Integrated Virus Detection System (IVDS).* ECBC-TR-811. US Army Edgewood Research, Development and Engineering Center: Aberdeen Proving Ground, MD.

Wick, C.H. and P.E. McCubbin. 2010. *Recovery of Virus Samples from Various Surfaces with the Integrated Virus Detection System.* ECBC-TR-816. US Army Edgewood Research, Development and Engineering Center: Aberdeen Proving Ground, MD.

Wick, C.H., P.E. McCubbin, and A. Birenzvige. 2005. *Detection and Identification of Viruses Using the Integrated Virus Detection System (IVDS).* ECBC-TR-463 (AD-A454 377). US Army Edgewood Research, Development and Engineering Center: Aberdeen Proving Ground, MD.

Wick, C.H., M.M Wade, T.D. Biggs et al. 2012. *Effects of Repeated Exposure to Filtered and Unfiltered Broadband Light Radiation on Escherichia coli Growth and Propagation.* ECBC-TR-987. US Army Edgewood Research, Development and Engineering Center: Aberdeen Proving Ground, MD.

Wick, C.H., S. Weugraitis, and P.E. McCubbin. 2010. *Integrated Virus Detection System Characterization of MS2 and TBSV after Pulsed Light Exposure.* ERBC-TR-817. US Army Edgewood Research, Development and Engineering Center: Aberdeen Proving Ground, MD.

Wick, C.H., H.R. Yeh, H.R. Carlon et al. 1997b. *Virus Detection: Limits and Strategies.* ERDEC-TR-453. US Army Edgewood Research, Development and Engineering Center: Aberdeen Proving Ground, MD.

Yates, J.R. 1998. Mass spectrometry and the age of the proteome. *Journal of Mass Spectrometry* 33: 1–19.

Index

Note: Page numbers ending in "f" refer to figures. Page numbers ending in "t" refer to tables.

Printed and bound by CPI Group (UK) Ltd, Croydon, CR0 4YY

24/10/2024

01778308-0005